Rain Catchment and Water Supply in Rural Africa: A Manual

Erik Nissen-Petersen

HODDER AND STOUGHTON

LONDON SYDNEY AUCKLAND TORONTO

in association with the Danish National Fund for Developing

British Library Cataloguing in Publication Data

Nissen-Petersen, Erik
Rain catchment and water supply in rural Africa.
1. Water conservation—Africa
I. Title
333.91'0960 HD1698.A/

ISBN 0 340 28429 3

First published 1982

Represented in East Africa by K. W. Martin, P.O. Box 30583, Nairobi, Kenya

Represented in Nigeria by Nigeria Publishers Services Ltd, P.O. Box 62, Ibadan, Nigeria

Printed in Great Britain for Hodder and Stoughton Educational, a division of Hodder and Stoughton Ltd, Mill Road, Dunton Green, Sevenoaks, Kent, by Richard Clay (The Chaucer Press) Ltd, Bungay, Suffolk

Contents

Acknowledgments

I am very thankful to the DANIDA (Danish International Development Agency) for its interest, and encouragement which have enabled me to experiment and test ideas over a period of 4–5 years.

Publication of this manual was also made possible by a Danish National Fund for Developing Countries grant and by DANIDA allowing me to spend working time on writing it.

Grateful thanks are also conveyed to Stan Stokoe and others who have scrutinised the text and made many useful suggestions.

Finally, but not least, I owe great thanks to my wife Wanza, for her patience and support without which it would have been almost unbearable to work in the harsh conditions of such an arid part of Kenya.

Conversion Tables

Millimetres to Inches

1 mm = 0.039 inches
2 mm = 0.078 inches
3 mm = 0.117 inches
4 mm = 0.156 inches
5 mm = 0.195 inches
6 mm = 0.234 inches
7 mm = 0.273 inches
8 mm = 0.312 inches
9 mm = 0.351 inches
10 mm = 0.394 inches

Centimetres to Inches

1 cm = 0.394 inches
2 cm = 0.788 inches
3 cm = 1.182 inches
4 cm = 1.576 inches
5 cm = 1.97 inches
6 cm = 2.364 inches
7 cm = 2.758 inches
8 cm = 3.152 inches
9 cm = 3.546 inches
10 cm = 3.94 inches
11 cm = 4.334 inches
12 cm = 4.728 inches
13 cm = 5.122 inches
14 cm = 5.516 inches
15 cm = 5.91 inches
16 cm = 6.304 inches
17 cm = 6.698 inches
18 cm = 7.092 inches
19 cm = 7.486 inches
20 cm = 7.88 inches
30 cm = 1 foot
61 cm = 2 feet
91 cm = 3 feet
100 cm = 39.37 inches
200 cm = 78.74 inches
300 cm = 118.1 inches

Area

1 sq m = 1.19 sq yds
1 sq km = 0.39 sq mile
1 ha = 2.5 acres
4050 sq m = 1 acre

Volume

1 cu m = 1000 litres
1 cu m = 220 gallons
1 cu m = 35.3 cu ft
1 litre = 0.22 gallons
4.55 litres = 1 gallon
1 oil barrel = 44 gallons
1 oil barrel = 200 litres

Volume to Weight

1 cu m water = 1.0 tonne
1 cu m sand = 1.7 tonne
1 cu m ballast = 1.6 tonne
20 bags of cement = 1.0 tonne
8 wheelbarrows sand = 1.0 tonne
1 oil barrel water = 0.2 tonne

Introduction

Modern water supply without pumps or pipes is possible when domestic water is situated next to your house. It may seem a little contradictory to say that water can be available at your house without some kind of piping, but if you have seen the rain on your roof and in your compound, you will have seen the great source from which all water originates; namely the rain.

Provided that it rains once or twice a year, it is possible to establish a reliable water supply system based on the catchment of this rain.

Naturally, many people will claim that it rains so little in their area, that no catchment system will ever be able to accumulate enough water for a whole year's consumption.

To these people I can say that my family and I have lived on three rain catchment systems for more than four years in the arid land of Makindu, in Kenya. Having only two rainy seasons, (and a rainy season can be as little as 100 mm – 4 inches), it gives two long dry spells in which demand for water is very high. But even during those dry periods we have been quite well-off for water.

One rain catchment system is a small earth dam built by hand. It holds enough water to grow sugar cane, bananas and citrus fruit on a 5-acre orchard irrigated by seepage only.

Another catchment system collects rain run-off from the compound around the house and diverts it to a storage tank in which we raise fish and ducks, as well as using the water for the production of biogas.

A roof catchment system, combined with a windpump and a solar panel, provide all the hot and cold water we want in the house throughout the year. In fact, during the last year my wife sold so much water that it paid back one quarter of the installation costs of our domestic water supply system.

Less complicated systems, which are still capable of supplying domestic water to a house, can be established for the cost of three goats. It seems, therefore, reasonable to relieve one's wife and daughters of the laborious work of carrying water long distances. The time and energy saved would allow them to work more on improving the family's welfare at home.

From an economic point of view, these catchment systems need no fuel, no spare parts and very little skill. They save your money as well as the country's foreign exchange.

These systems do not permit the breeding or spreading of those diseases which are often found in common waterholes and other water-sources used in rural areas.

Furthermore, these low-cost systems, when used appropriately, promote rural employment, skill and self-reliance.

The only excuse for people not doing something about their water problems at home, is ignorance of what they are able to do themselves.

This manual is produced as an attempt to open the eyes and minds of people as well as to help them to find a practical solution to their water problems.

The manual is not a complete list of all possible solutions to water problems in semi-arid and arid lands but it is hoped that it will create some rethinking on the subject and that many new ideas will surface.

The author will be only too glad to hear your comments and your ideas.

Finally, it is hoped that this manual will be a valuable contribution to the *International Drinking Water Supply and Sanitation Decade, 1981–1990*, sponsored by the United Nations Development Programme.

1 Water Requirements

Three types of water for a homestead

First it is important to determine two things: (i) for which purposes do you need water, and (ii) how much water do you need for these purposes.

For a homestead in a rural area use of water may be divided into three groups:

a) clean water for domestic use in the house,
b) fairly clean water for livestock, poultry, etc, and
c) unclean water for irrigation of crops.

Clean water can, of course, be used for all purposes, but since it is the most difficult water to collect and store, it is not reasonable to use it where more easily accessible water can be utilised.

Unclean water can be treated for domestic use, but since it involves two extra operations – filtering and transporting it to your house – it should not be defined as clean water in this connection.

Consumption rates

Domestic water

When calculating the required volume of clean water for domestic usage you have to decide which level of installation you want to design for your water supply. Table 1 indicates five levels of which the middle one, having water next to house, will satisfy the majority of peoples' needs.

Table 1 Domestic consumption of water by 1 person.

Living 15 km from a water source	2–3 litres daily
Living 1 km from a water source	3–6 litres daily
Having water next to house	10–20 litres daily

Having a watertap, shower and adjusted WC	60–80 litres daily
Having full sanitary installations	175–250 litres daily

Having decided which rate of consumption per person you want, then fill in the figure in Table 2. Also fill in the number of persons living in your house and the number of days of the longest period without rain.

Multiply the figures and you will reach the volume of water required for domestic use, which is also the minimum size of storage tank to be built.

Table 2 Calculation formula for domestic water.

Family members living in house. No. of persons.	Average daily consumption. Litres per person	Longest period without rain No. of days.	Volume of water required to last through dry season
. . . . . persons	× . . . . . litres	× . . . . . days	= . . . . . litres
Example:			
8 persons	× 15 litres	× 180 days	= 21,6000 litres.

Water for livestock

To estimate the volume of water needed for your livestock fill in Table 3.

Table 3 Calculation formula for livestock water.

Type and No. of livestock	Daily consumption rates per animal. Litres per animal.	Longest period without water No. of days	Volume of water required to last through dry season
. . . . . upgraded dairy cattle	× 100 litres	× . . . . . days	= . . . . . litres
. . . . . upgraded beef cattle	× 50 litres	× . . . . . days	= . . . . . litres
. . . . . local cattle	× 20 litres	× . . . . . days	= . . . . . litres
. . . . . sheep	× 5 litres	× . . . . . days	= . . . . . litres
. . . . . goats	× 3 litres	× . . . . . days	= . . . . . litres
Poultry, dipping, biogas, etc.		× . . . . . days	= . . . . . litres
		TOTAL	. . . . . litres

Example:

10 local cattle	× 20 litres	× 120 days	= 24,000 litres
60 goats	× 3 litres	× 120 days	= 21,600 litres
Poultry, dipping, biogas, etc.		× 120 days	= 3,400 litres
		Total	49,000 litres

If water for dipping livestock is to be drawn from the livestock water reservoir, then add 3 litres per head of livestock per week to the estimated volume of the reservoir.

Fish can easily be raised and fed on chicken manure in the reservoir without needing any additional volume of water. Chickens, pigeons and turkeys can live on used water from the house, so it is not necessary to calculate any additional water for them, but ducks and geese need about 1 litre of water daily per bird.

If you have production of biogas in mind, then a weekly consumption of about 20 litres of water must be included in the total requirement of water for livestock.

Your water requirements

By now your two types of water requirements should be known and it will be useful later on to have the figures filled-in, as in Table 4.

Table 4a Total requirements for water consumption.
.... litres of clean water for domestic use.
.... litres of fairly clean water for livestock, etc.

Example

Table 4b Total requirements for water consumption.
21,600 litres of clean water for domestic use.
49,000 litres of fairly clean water for livestock, etc.

Your requirements of water are now known. The next problem is to find a source of water which can fulfil your needs.

The next two chapters will deal with this interesting problem.

2 The Source of Water

The water's cycle

All water on the earth's surface is moved in large circular patterns by the sun and the force of gravity.

From the sea and land the sun evaporates the water, and forms clouds. Differences in temperatures, created by the sun also create differences in pressure which generate the wind to blow the clouds to other regions where the moisture will fall as rain or snow.

There are four things which can happen to rain when it falls on land:

a) it flows to lower areas and ends up in seas or lakes – a process called run-off,
b) it seeps into the ground and slowly moves towards the sea – a process called ground-water flow,
c) it is sucked up by vegetation and evaporates – a process called transpiration,
d) it evaporates directly from the surface.

In all cases, sooner or later the water will evaporate somewhere and the cycle goes on, as shown in Fig. 1.

All water, even that being pumped up from deep boreholes, originates from rain. When rain falls it is clean water, but during its run-off some of it gets dirty or salty and has therefore to be treated before it can be used.

By catching rain when it comes straight from the clouds we can get clean water, and if it is collected at a higher level than the place of consumption, eg on the roof, the water will flow by its own gravity to your watertap.

Therefore collecting rain from your roof will provide you with free clean water delivered to your door.

You do not need filters, pumps or pipes.

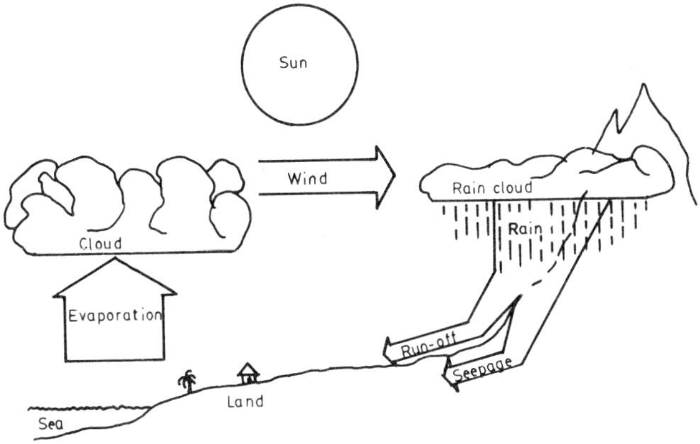

Fig. 1 Water Cycle

Rainfall patterns in Kenya

Before going into details about rain catchment, it might be useful to understand some of the mechanics of the rains.

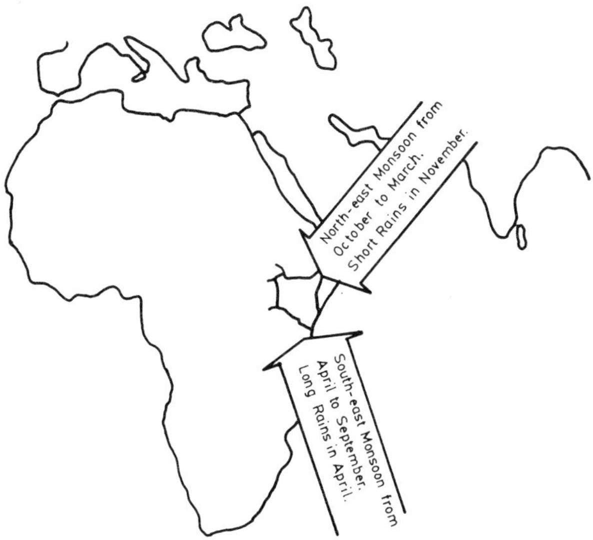

Fig. 2 The Two Monsoons Affecting Kenya

The weather pattern over Eastern Kenya is mainly dominated by two monsoons, the south-east and the north-east. They are controlled by a large-scale pressure system over the western Indian Ocean. Other areas to the west of the Rift Valley only experience one rainy and one dry season a year.

The south-east monsoon starts to blow from the southern Indian Ocean in March–April. When it meets air masses of the north-east monsoon over East Africa, it results in the so-called Long Rains. The south-east monsoon takes control of the area and sends dry, dull air from the southern part of Africa until September. Then the north-east monsoon regains power and with hot, dry air from the Horn of Africa it forces back the south-east monsoon. The battle produces the Short Rains in November, and the south-east monsoon has to retreat until March/April at which time the story repeats itself.

Depending to a certain extent on the situation of the monsoons' battlefields, the rains fall in various patterns over Kenya. The areas with most reliable rainfalls are situated near Mount Kenya and Lake Victoria, while the semi-arid and arid zones of the lowland such as Makindu, receive more unstable rainfall, as illustrated in Fig. 3.

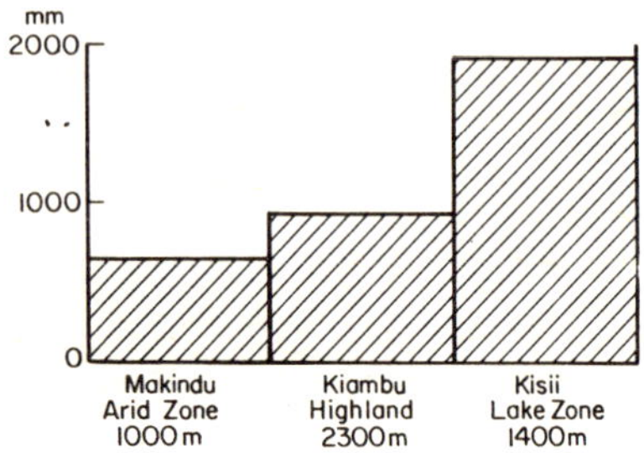

Fig. 3 Average Annual Rainfall

Monthly fluctuation of rainfall

While the rains vary from year to year, they also vary from month to month, as shown in Fig. 4. Again we can see that the semi-arid and arid zones, such as Makindu, have little rainfall and a long, dry season with virtually no rain at all.

Despite this unfortunate situation, it is still possible to construct reliable water systems based on rain catchment. Just remember the results described in the introduction to this manual.

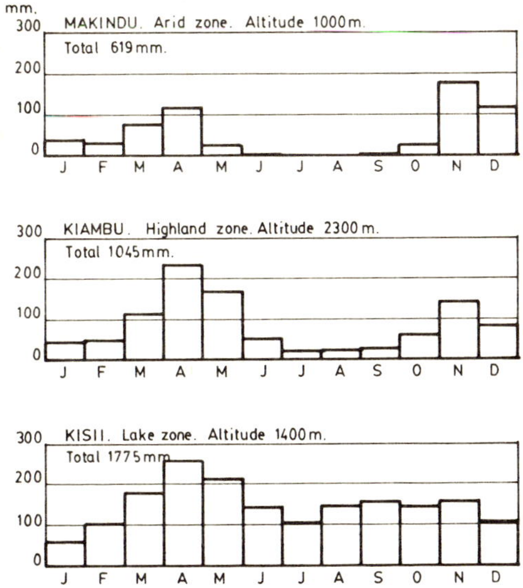

Fig. 4 Average Monthly Rainfall

Rainfall patterns in Nigeria

Nigeria has a tropical climate which is hot and wet throughout the year in the south-east, but has a marked dry season in the west and north, which increases inland from the coast.

Two air masses, the Equatorial Maritime air mass bringing rain from the Atlantic to the south, and the Tropical Continental air mass bringing hot, dry winds from the Sahara to the north, dominate the weather system. Thus the length of the rainy season decreases from nine months in the south (March to November) to only four and a half months in the north (May to September), and the harmattan, a dry, dusty wind from the Sahara, blows for three months in the north but for only a couple of weeks in the coastal belt. Figs. 5 and 6 show these air masses and the resulting annual rainfall pattern in three regions of Nigeria.

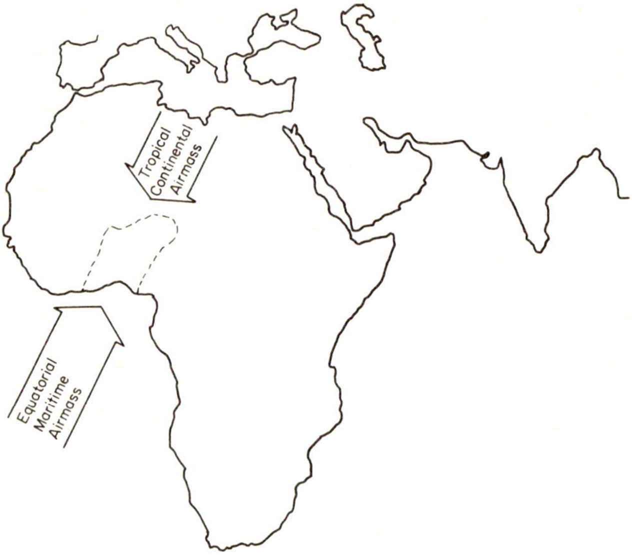

Fig. 5 Air Masses Affecting Nigeria

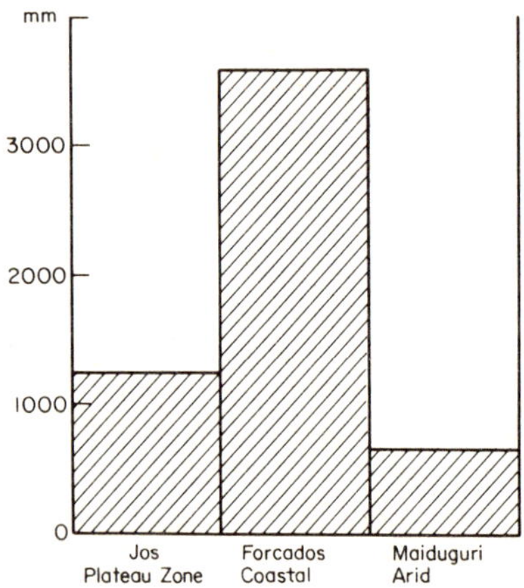

Fig. 6 Average Annual Rainfall

Monthly fluctuation of rainfall

The rainfall varies from month to month, depending on the season. Fig. 7 shows these fluctuations in three areas – a highland zone, a dry northern region and a wet coastal region.

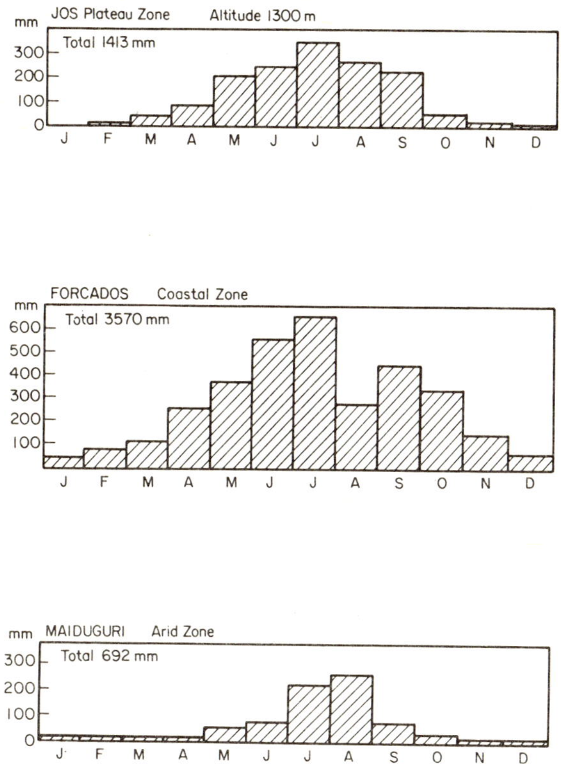

Fig. 7 Average Monthly Rainfall

Rainfall patterns in Zambia

Zambia, too, has a tropical climate with a short rainy season lasting from November to March or April and then little or no rain for the rest of the year.

There are two airstreams which affect the Zambian climate. Humid air, which originates over the South Atlantic Ocean, brings rain during the wet season to the northwest of the country. The northeast monsoon originates over Asia, crosses the northern Indian Ocean and may reach northeastern Zambia in midsummer. The air contains a great deal of moisture after its long journey across the sea, so heavy rains often result. Tropical cyclones may also bring very heavy rain to parts of the country.

Fig. 8 below shows the annual rainfall pattern in three regions of Zambia.

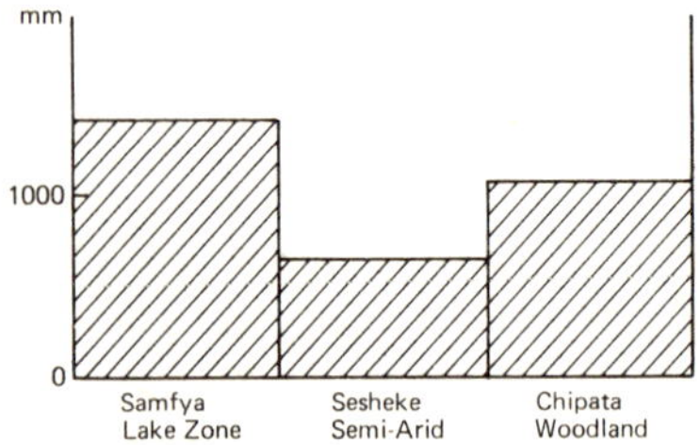

Fig. 8 Average Annual Rainfall

Monthly fluctuation of rainfall

Over most of the country there is a single rainfall peak, the rainiest month being January. However, in the north there is a tendency towards two peaks of rainfall in December and March.

Rainfall patterns in the Sudan

Lying completely within the tropics and far from the sea – the effect of the Red Sea to the east on the rainfall system is negligible – the country is very hot, especially in the central regions. Southwards from the deserts the rainfall increases slightly until a fairly dependable supply is found in the south of the country where the savannah changes to forest in the mountains.

The level of rainfall fluctuates from month to month depending on the region. The deserts tend to have one heavy fall a year, whereas the rest of the country has two rainy seasons, as is typical in the tropics.

Fig. 9 below shows the annual rainfall for three different regions of the Sudan.

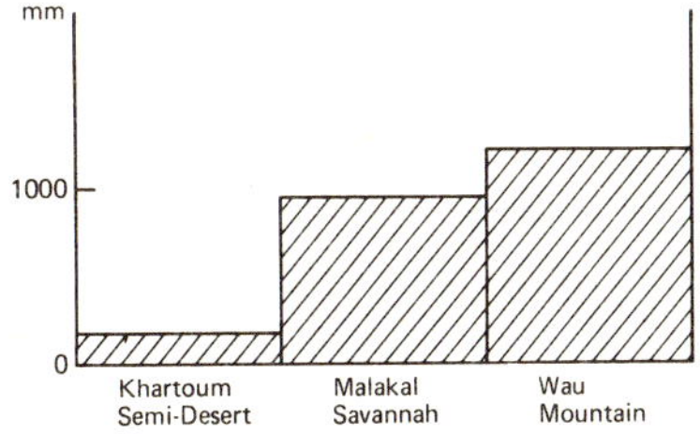

Fig. 9 Average Annual Rainfall

Rainfall patterns in the Somali Republic

The climate of the Somali Republic is hot and despite the country being situated near the sea rainfall is very limited in many places. It is only really in the mountains in the north that a moderate fall occurs.

There are two rainy seasons in the year, the amounts varying from region to region.

3 Catchment Areas

Three types of catchment areas

Rain catchment depends on two things: (i) rain, and (ii) the area on which the rain falls. These areas are called catchment areas and may be divided into three types depending on the surfaces of the catchment areas.

1st type: *Total Run-off* occurs in areas with hard surfaces such as roofs and rocks.

2nd type: *Half Run-off* comes from semi-hard surfaces eg roads, compounds around a house and rocky slopes.

3rd type: *Quarter Run-off* may be collected from catchment areas with loose soil surfaces such as fields and valleys.

These types of run-offs correspond with those three groups of water mentioned in Chapter 1 in the following ways:

Total Run-off from roofs and rocks gives clean water for domestic use.

Half Run-off from roads and compounds provides fairly clean water for livestock, fish farming and biogas.

Quarter Run-off from fields and valleys is suitable for irrigation, construction of dams or sinking of wells.

Total run-off for domestic use

Collecting rain from the roof of one's house should be a must for every houseowner, because it can provide clean water for one's family throughout the year.

To calculate the volume of water available from your roof do as follows: First measure the length and the width of the roof in metres, then multiply the two figures and you will know the roof area in square metres (sq m).

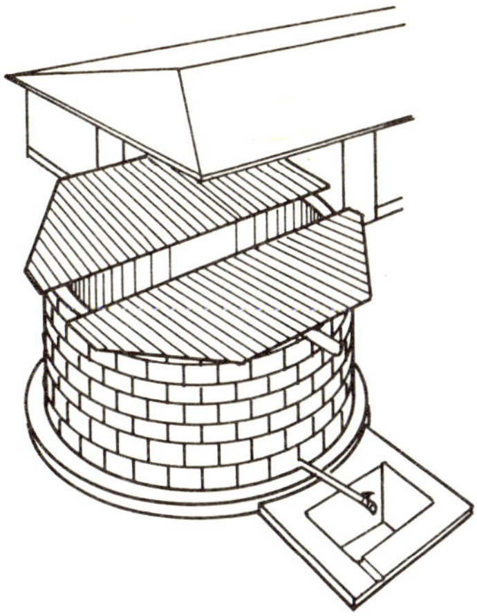

Fig. 10 Roof Catchment

Thereafter ask your local agricultural adviser or meteorological officer for the mean annual rainfall in millimetres (mm) at your place. Then multiply this annual rainfall figure with the sq m of your roof and you will get the average annual volume of water in cubic metres (cu m) which can be collected from your roofs.

Table 5 Calculation of Total Run-off from a Roof.

Size of Roof. sq m	Rainfall. mm	Run-off. Cu m (tonnes).
. . . . . sq m	× . . . . . mm	= . . . . . cu m (tonnes)
Example: Makindu region, Kenya.		
60 sq m	× 619 mm	= 37,140 cu m (tonnes)

The example in Table 5 shows that from a roof of ordinary size in a dry region you can collect 37 tonnes of clean water in an average year. That can provide 7 persons with 15 litres of clean water every day in the year.

If more water is required for domestic use the roof can be extended, a rock catchment included or water can be drawn from one of the other catchment systems mentioned below.

In areas with rocks or mountains it is possible to catch rain water from these structures. It can be done either from the rock itself or the water can be fed to a storage tank situated nearer the house.

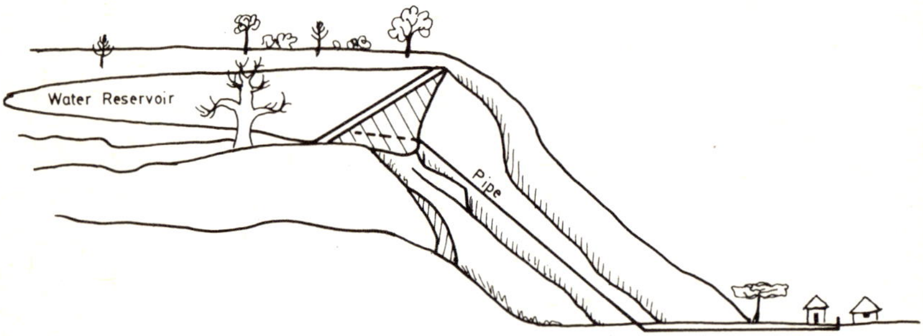

Fig. 11 Rock Catchment

The advantages of Total Run-off systems are that even a small rain shower will supply clean water, and that the run-off is easy to store in a tank situated next to the house.

Half run-off for livestock, fish farming and biogas
As the name of this type of catchment indicates, you can collect about half of the annual run-off from, for example, the compound surrounding your house or from the road leading to your house.

If your catchment area is well-sloped and hard-surfaced you can collect as much as three-quarters of the annual rain, but if it has little slope and a porous surface you may only catch about one quarter of the rain.

Table 6 Calculation of Half-Run-Off from a Compound.

Size of area. sq m	Rainfall. mm	Surface. hard – ¼ medium – ½ soft – ¾	Run-off cu m (tonnes).
. . . . . sq m	× . . . . . mm	–	= . . . . . cu m
Example: Makindu region, Kenya.			
200 sq m	× 619 mm	– ½	= 61,900 cu m

The run-off is led by way of a ditch to a storage tank dug into the ground. This tank can either be a waterhole with a clayish surface or, more permanently, a tank made out of mortar reinforced with chicken wire (ferrocement).

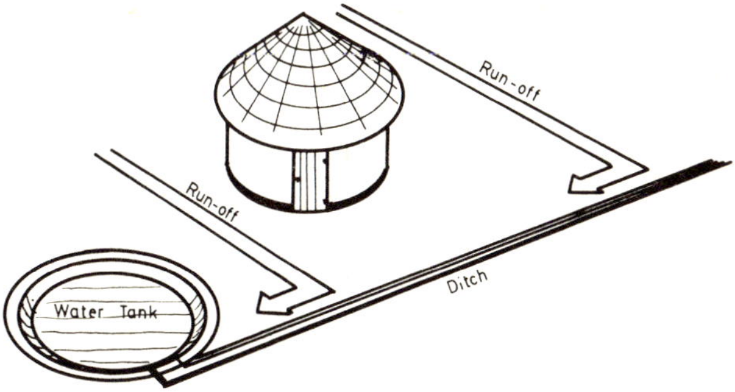

Fig. 12 Compound Catchment

To compensate for little slope, soft surface or a need for more water, the catchment can either be extended or covered with a hard surface, eg clay or laterite (murram).

Build-it-yourself instructions and drawings are to be found on page 57.

Quarter run-off for irrigation and construction of dams

For several reasons this type of rain catchment is not as easy to handle as the previous two types, but if we leave out construction of earth surface dams and concentrate on subsurface dams and wells, we can keep the subject within the do-it-yourself capability.

For those who would like to consider building a surface earth dam, a study of Mr N W Hudson's book, 'Field Engineering for Agricultural Development', Oxford University Press, is recommended. It describes, among other things, the importance and guidelines for the construction of a proper spillway design which is the most essential factor in designing a long-lasting dam.

Subsurface dams

There are three ways of constructing subsurface dams:

a) The easiest and cheapest method is to construct a clay or other non-porous barrier across the riverbed, which rests on the solid bedrock of the watercourse (see Fig. 13).

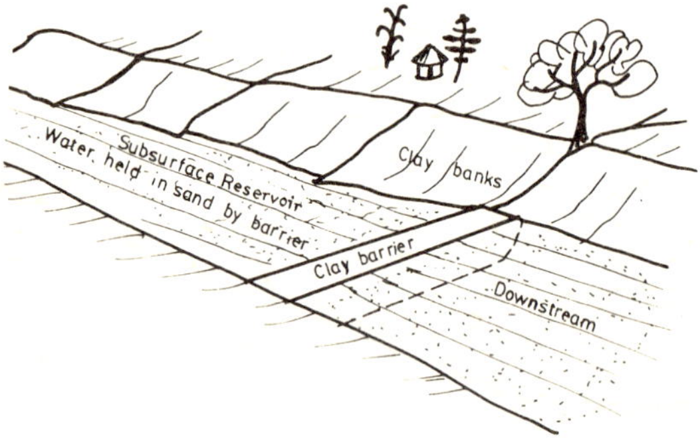

Fig. 13 Earth Subsurface Dam

b) The second construction is a gravity concrete dam or weir which is situated across the watercourse and also reaches to the non-porous bed-rock. It protrudes above the riverbed to catch more water, thereby increasing the volume of the reservoir (see Fig. 14).

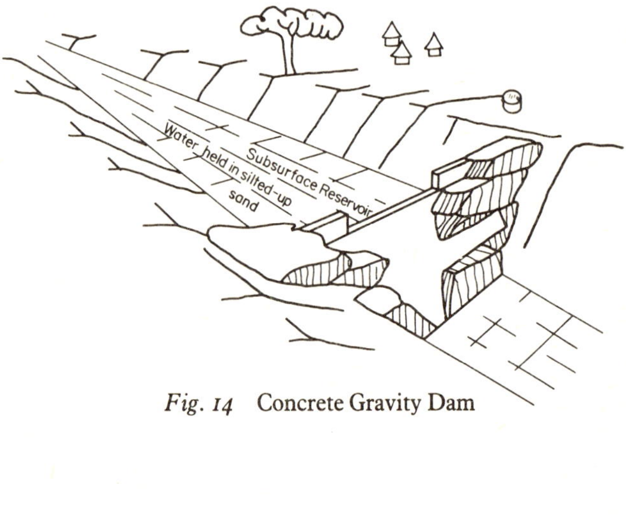

Fig. 14 Concrete Gravity Dam

Building instructions for the concrete gravity dam are to be found on page 63.

c) The third construction is an arch weir built of blocks or concrete as a thin walled arch across the riverbed. This type of subsurface dam is fairly cheap to build, but it requires rock abutments on both river banks (see Fig. 15).

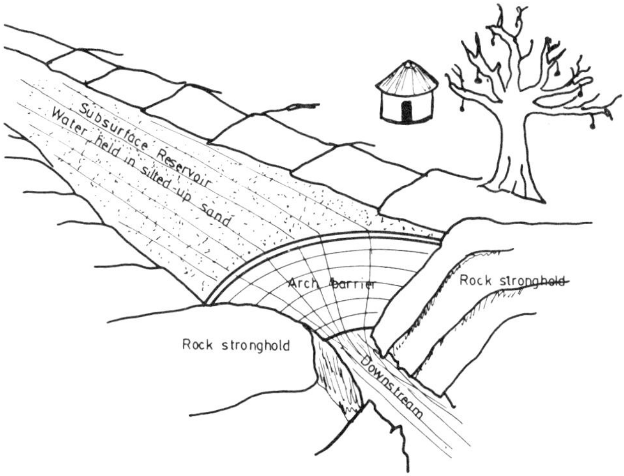

Fig. 15 Concrete Arch Dam

These dam types are easy to design and construct because surplus run-off just passes over the top of the dam without damaging it, thus no spillway design is necessary. Other advantages are:

a) As part of the dam structure the reservoir is silted up by coarse sand. Thus silting is not a problem but a gain.

b) Water is stored between the coarse sand, so evaporation is eliminated to almost zero.

c) Mosquitoes and bilharzia parasites cannot breed in a subsurface reservoir. Therefore there will be no increase in malaria and bilharzia.

d) Because people, livestock and wild animals cannot see the water, it is easier to prevent contamination.

e) By drawing water from the reservoir through a well situated either upstream or downstream of the dam structure, it is not difficult to maintain a clean and reliable water source.

4 Storage Tanks and Reservoirs

Determine capacity of tanks

In the previous chapters you have determined the volume of water needed and where to get it from. In this chapter we shall deal with the storage of water.

The first thing to decide is the capacity of the tanks required for holding enough water from one rainy season, through the drought, to the next rainy season. Using the example from the dry region again, we can see from Table 4 that we need a tank with a minimum volume of 21,600 litres for domestic water and a 49,000 litre tank for livestock, fish-farming and the production of biogas.

Example from page 00

Table 4c Total requirements for water consumption.
21,600 litres of clean water for domestic usage.
49,000 litres for fairly clean water for livestock, etc.

Seepage from a concrete subsurface dam with a volume of 40,000 cu m is sufficient to irrigate an area of 1 ha (2.5 acres). It must be noted, however, that this seepage will not prove effective for shallow-rooted crops, but is totally adequate for tree and other long-rooted crops.

Calculation of tank and reservoir volumes

Roof catchment tank
The strongest and cheapest tank for roof catchment is a cylinder with a

diameter greater than its height. The height is always determined by the distance between the surface of the tank foundation and the lowest point of the guttering. To check the lowest point pour water into the guttering or if no guttering is in place yet, then use a hose pipe filled with water fixed onto the place of the planned guttering.

The formula for calculating the volume of a cylindrical tank is as follows:

$$\frac{\text{Radius} \times \text{Radius} \times 22 \times \text{Height}}{7}$$

(Diameter is the longest distance possible across the floor of a tank. The Radius is half that of the Diameter and is measured from the centre of the tank to its side.)

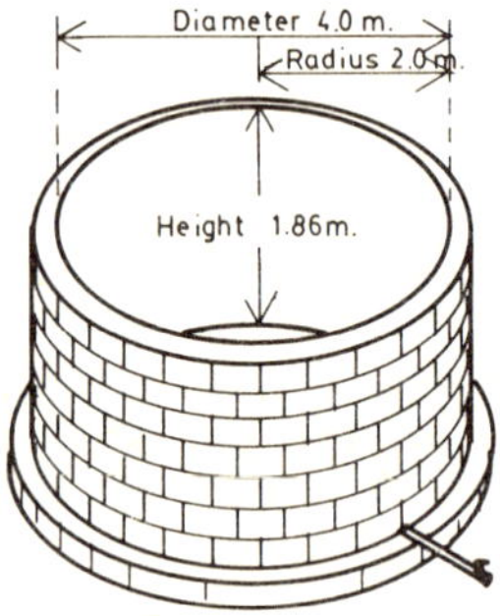

Fig. 16 Roof Catchment Tank

Therefore the measurements for a 23,383 litre tank with height of 1.86 m will be:

$$\frac{\text{Radius } 2.0\text{ m} \times \text{Radius } 2.0\text{ m} \times 22 \times \text{Height } 1.86\text{ m}}{7} = 23.383 \text{ cu m}$$

In case you want another size for your tank, then insert other figures for height and radius and try it out until you find a combination which suits your purpose.

Compound catchment tank

The most economical shape and construction of a compound catchment tank is that of half a sphere dug into the ground.

The measurements for a 51,000 litre tank is according to the

formula
$$\frac{4 \times 22 \times R \times R \times R}{3 \times 7 \times 2} = \text{Volume}$$
as follows:
$$\frac{88 \times \text{Radius } 2.9\text{ m} \times \text{Radius } 2.9\text{ m} \times \text{Radius } 2.9\text{ m}}{42} = 51{,}100 \text{ cu m}$$

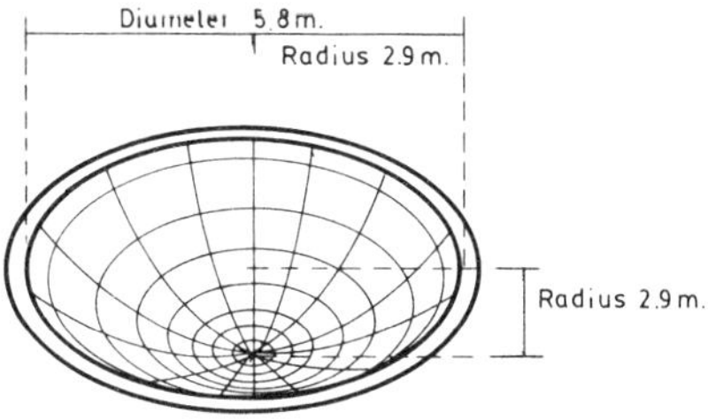

Fig. 17 Half Sphere Compound Tank

Again, if you want a bigger or smaller tank then insert a bigger or smaller figure for the radius and try it out until you reach the tank size you want.

Subsurface reservoir
Due to several unknown factors, eg the amount of sand, the angle of the sides of the reservoir and the permeability of the adjacent soil, it is only possible to estimate an approximate volume of a subsurface dam.

The formula is
$$\frac{L \times W \times H}{6} = \text{approx volume.}$$

A subsurface dam of 8,000 cu m capacity needs the following design L = 550 m, W = 35 m, H = 2.5 m as
$$\frac{550 \times 35 \times 2.5}{6} = 8{,}000 \text{ cu m.}$$

Calibration of tanks

To enable you to keep an eye on the level of the water, it is useful to paint a graduation scale of volume on the side of your tank wall. This calibration can be carried out either through calculation or

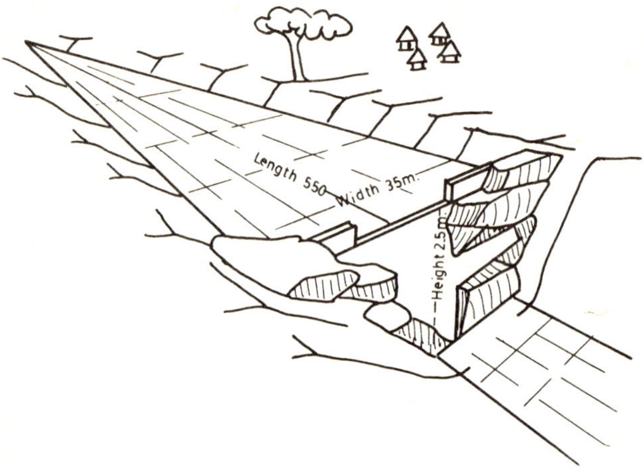

Fig. 18 Subsurface Dam

by putting a known volume, eg oil barrels, each containing 200 litres of water, into the tank until it is full. For every 1,000 litres poured into the tank a mark and the number of 1,000 litres should be marked on the wall with oil paint.

The calibration formula for a cylindric tank is:

$$\frac{\text{height of tank in cms}}{\text{volume of tank in cu m}} = \text{height in cms per 1,000 litres.}$$

To scale the 23,383 litre tank mentioned above:

$$\frac{186 \text{ cm}}{23.383 \text{ cu m}} = 7.96 \text{ cm} = 8 \text{ cm height per 1,000 litres.}$$

The calculation formulae for half sphere and jar shaped tanks is complicated, but if you want to calibrate them, then work them out by filling them with a known volume and marking them with oil paint.

Size and situation of tanks

Looking at the former examples from the dry region, we said that a tank capacity of 23,383 litres was needed and we know from Table 5

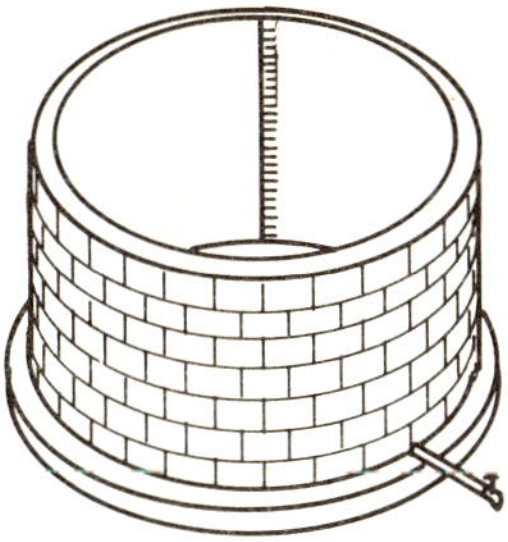

Fig. 19 Calibration Scale

that a roof area of 60 sq m in Makindu is required for filling that tank's volume. In case the roof area of your house is less than that, you might include one or two roofs of the other houses you have got on the compound. If the latter is feasible, then you have to decide whether it is practical to lead all run-off into a 23,383 litre tank or whether it is more suitable to build a smaller tank next to each house.

One large tank is usually cheaper than two or three smaller ones, but the cost depends mainly on the type of tank you prefer. Mentioned below are six different types of tanks which are easy to build and install.

Corrugated iron sheet tank

The quickest and easiest way of establishing a roof catchment system is to buy and install corrugated iron sheet tanks but these tanks have some drawbacks. Iron sheets are affected by rough handling and transportation, and after some years leakage will occur due to corrosion and other factors.

These drawbacks can be prevented by coating a new tank with bitumen all over and then placing it on a reinforced concrete foundation with an extension for a ferrocement coating of the outside wall of the tank. These precautions will, together with the cost of purchasing and transporation, make it an expensive enterprise and the time saving aspect will also be minimal.

Another drawback is the limitation of tank size. Water is heavy, 1 tonne per cubic metre, so iron sheet and especially its jointing and soldering have a certain strength limitation which it is risky to exceed.

Concrete ring tank

To avoid the disadvantages of corrugated iron sheets, and to maintain a single and easy construction procedure, concrete septic tank sections can be built to form water tanks 2 cu m in volume. The volume of the small tanks is suitable for rain catchment from small roofs scattered on a compound as well as for areas with monthly rainfall.

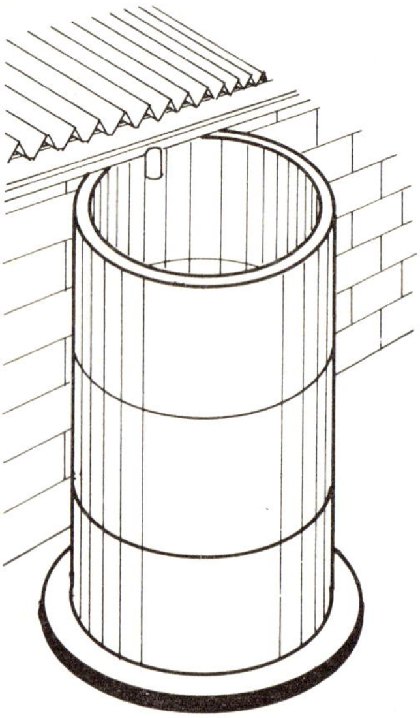

Fig. 20 Concrete Ring Tank

The cost of purchasing and transporting the tank sections can be greatly reduced by concreting the rings yourself, in a steel formwork. This formwork is, however, rather expensive for individuals, but quite reasonable for a community or a development agency.

Concrete block tank

Storage of 23,000 litres of water in one tank requires cement or quarry blocks of a quality stronger than blocks for a house. If you would like to control the quality and at the same time reduce the cost of purchasing and transporting blocks, you can make your own blocks.

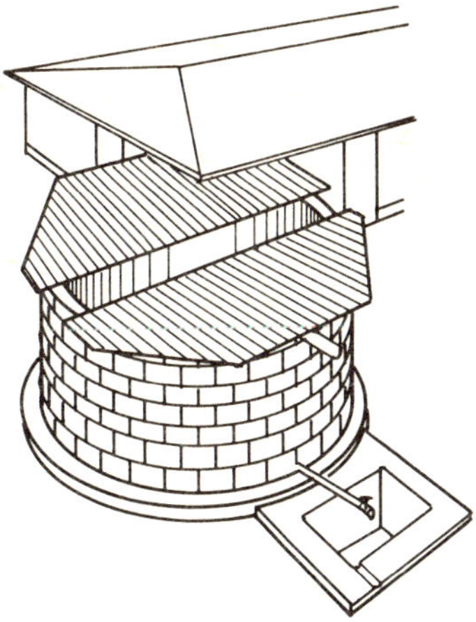

Fig. 21 Concrete Block Tank

The site for a tank of this size has to consist of firm soil which has never been dug out, for instance places where soil has been previously dug out for plastering a house or a rubbish pit. Also new or old tree roots can lead to cracks in the foundation and tank wall.

Build-it-yourself instructions and drawings are to be found on page 42.

Granary basket tank (UNICEF design)
This tank type uses a granary basket of woven sticks as a built-in framework for a ferrocement tank. The cost of the framework is thus reduced to a few days work in cutting and weaving sticks into an open weaved basket.

Using a method called ferrocement, it is possible to improve the strength and increase the size of the tank by covering the outside of the basket with a layer of chicken wire rounded up with barbed wire before plastering the basket inside as well as outside.

Without wire reinforcement the tank size should not exceed 3.5 cu m but using the ferrocement technique it is possible to triple the volume and the life of a tank.

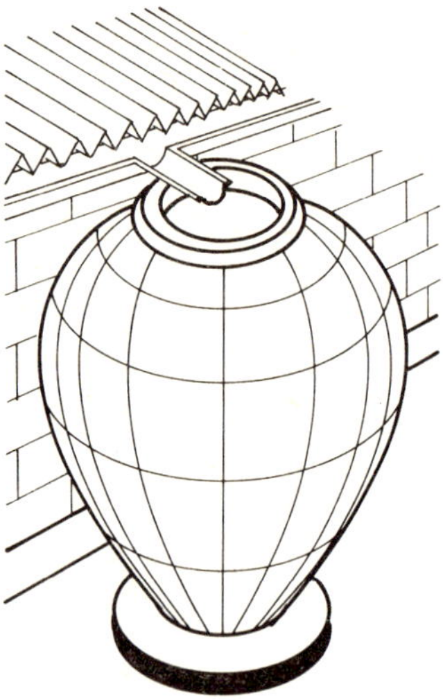

Fig. 22 Granary Basket Tank

Large cement jar (UNICEF design)
The framework for this tank is a large bag made of cloth or sacking and stuffed with grass. Mortar is plastered onto the bag, chicken and barbed wire are tied onto the plaster and the whole thing is plastered once more. 24 hours later the bag is removed from the inside of the jar and it is plastered to make it waterproof. The bag can be used for many water jars, so the cost per tank is minimal.

Tank volumes can be as great as 10 cu m. Besides being designed for water storage, they can also be used for storing grain.

Small cement jars (UNICEF design)
These are similar to the above mentioned cement jars except for their smaller size and lack of reinforcement. They can store 1 cu m of water or grain and they cost as little as a bag of cement each.

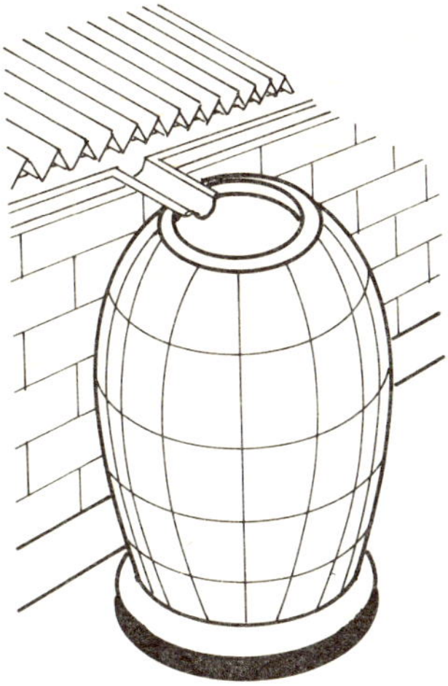

Fig. 23 Large Cement Jar

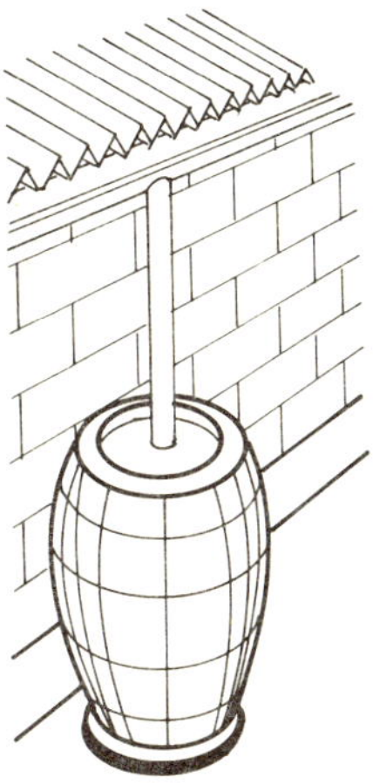

Fig. 24 Small Cement Jar

Guttering

Also worth considering is the type of gutters you would like to use. Iron sheet gutters are the best, but unfortunately also the most expensive.

To save costs in the initial stages of constructing your water supply system, it may be wise to use cheap local materials for a while.

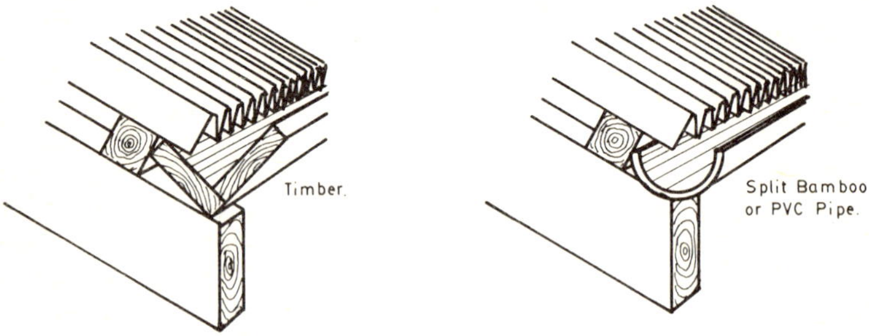

Fig. 25 Alternative Guttering

5 Wells

Definition of wells, boreholes and tubewells

There is some confusion about the use of the words wells, boreholes and tubewells, so let us get these terms straight first.

In English a borehole is a drilled hole with a small diameter and it can be several kilometres deep. A well is a hole dug with a diameter big enough to allow a man to work in and with a maximum depth of about 75 metres. A tubewell is a perforated pipe with a pointed end which is hammered into the ground.

When a well is less than 10 metres deep we call it a shallow well and when more than 10 metres deep, it is called a deep well. Also the various methods of constructing wells have got their special names: an earth well is just a hole in the ground, a masonry well is built of bricks, and a sinking well is constructed and then sunk in stages from the ground level.

Location of well sites

If you are not a geologist or someone with special knowledge, it is difficult to know where water can be found a short distance underground. There are, however, some places where it is worthwhile to start test digging:-

a) Upstream and downstream of dam reservoirs.
b) In sandy riverbeads, especially upstream of bedrocks.
c) In foothills of mountains or big rocks. Look for green trees and holes dug by animals.
d) In areas with green vegetation during drought.
e) Near existing wells and waterholes.

Having found a potential site, then dig as deep down as you can.

Remember that loose soil can collapse on you, so don't excavate the sides of the hole too steeply. It is possible that you will hit a rock which has to be removed in small pieces. A small rock can be broken up by a sledge hammer, and a bigger one by heating it with a charcoal fire and then cooling it down quickly by pouring a drum of water on to the heated rock. The rapid temperature changes will crack it into small pieces.

If you can get hold of an auger hand-drill, it will save you a lot of excavation work, because drilling in the bottom of a hole will tell you about the ground further down. To ease the work with the auger drill pour water into the drilling holes to soften the soil.

The first hole might be a wasted effort, but don't let that discourage you because by now you have learned a lot about siting, soil structure and digging methods. The next hole will certainly be more successful as well as easier to dig.

Well types

There are several ways and methods of constructing wells, all depending on the purpose of the well, soil structure, water source and local skills. This manual will only mention four types of well, so if you would like to learn more on the subject, it is recommended that you should read some of the publications listed in the bibliography.

Oil barrel well

To start with the construction of a well that everybody can afford and build himself, we have the oil barrel well. All you need is a hole in the ground with water and some worn-out, perforated oil barrels. Cut the tops and bottoms out of the barrels and clean the remaining cylinders. Place the first barrel cylinder in the bottom of the well and put stones or gravel around it. Place the next barrel on top of the first one, fill around it, and then place the next barrel and so on until you have reached about half a meter above ground level.

Ferrocement well

A shallow earth well can be made safer and probably deeper by ferrocementing the wall of the well and by building a well head above ground level.

The wall of the earth well is excavated to a straight and smooth

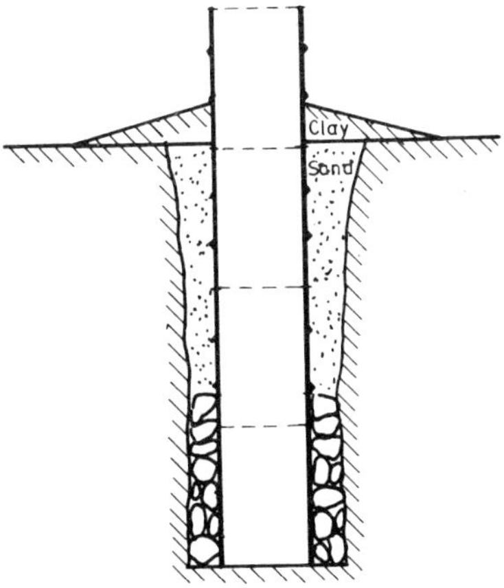

Fig. 26 Oil Barrel Well

surface which is then plastered with a layer of mortar, then reinforced with a layer of chicken wire, and finally plastered a second time.

This construction technique is called ferrocement.

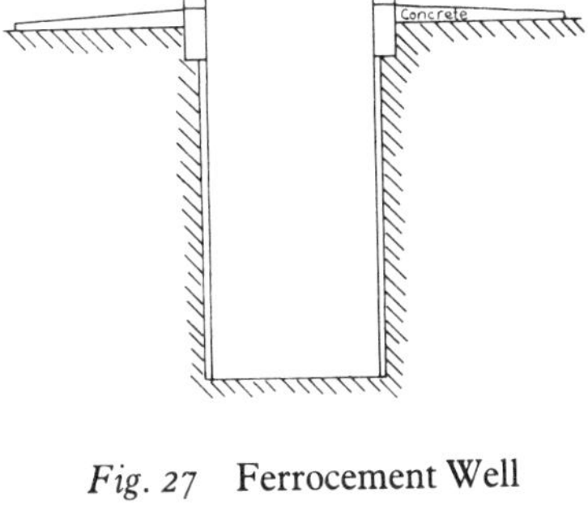

Fig. 27 Ferrocement Well

A so-called 'well head' is built above ground level to lessen the risk of children and animals falling into the well.

To prevent contamination of the well a concrete apron, sloping away from the well head, is laid out on the surrounding soil.

Concrete ring well (Sinking well)
This method requires either a steel cassion ring mould for concreting rings on the site, or the rings to be bought and transported from a factory to the construction site. Both alternatives are expensive for an individual, but practicable when a number of wells are being made. (Read *Shallow Wells*, Forth Progress Report, P.O. Box 85, Amersfoort, The Netherlands.)

The rings, measuring a height of 1.0 metres and a diameter of 1.2 metres, are stacked upon each other in an excavated well hole or they can be used for sinking wells or a combination of both procedures can be used.

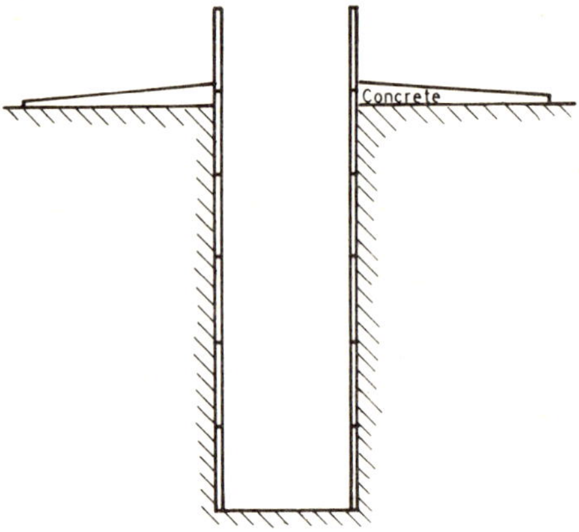

Fig. 28 Concrete Ring Well (Sinking Well)

This and the following type of well are also called 'sinking wells', because they can be sunk safely through sandy soil.

By digging soil out from the bottom of the well, the whole structure will sink. When the top of the well has reached the surface of the surrounding soil a section is added onto the top of the well. Thereafter digging is repeated until another section can be added onto the well at ground-level and so on until a satisfactory depth is reached.

Concrete block well (Sinking well)
Instead of using concrete rings, it is handier and cheaper to use concrete blocks shaped in a wooden form. These blocks are built onto a foundation concrete ring, which can be formed in a wooden shuttering, or more cheaply, in a ditch in the soil at the construction site.

Build-it-yourself instructions and drawings are to be found on page 55.

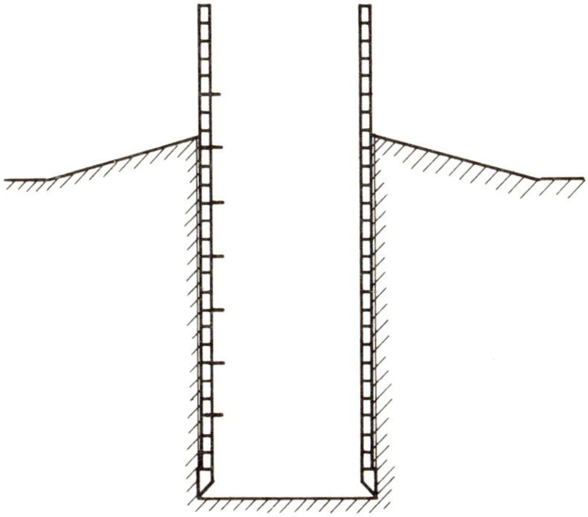

Fig. 29 Concrete Block Well (Sinking Well)

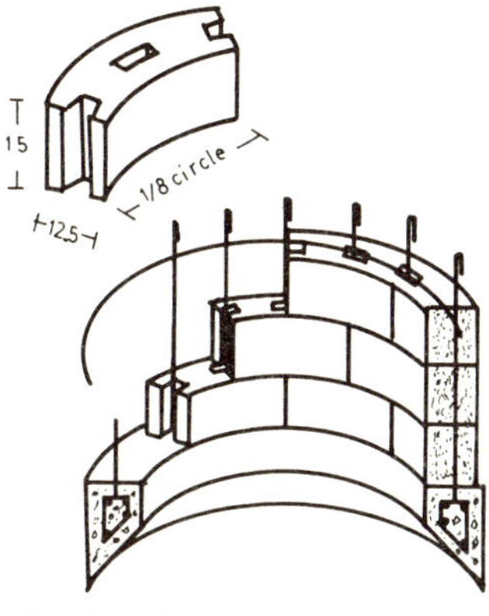

Fig. 30 Section of Foundation and Reinforcement

Aspects of hand-pumps

According to experience and with reference to an Oxfam paper *Hand-pump Maintenance*, published by Intermediate Technology, London, 'as many as 40 to 80 per cent of hand-pumps are inoperative within 3 years of their installation.' Also quoted from the booklet: 'Anybody who cannot afford or cannot maintain a factory-made pump is generally better advised to resort to a bucket and rope, with a windlass for raising and lowering'.

Simple lifts for walls

Calabash lift on roller
The simplest kind of locally made lift is a calabash on a sisal rope being pulled over a wooden roller. Fix a stone in the handle of the calabash, so that it easily tips over into the water in the well.

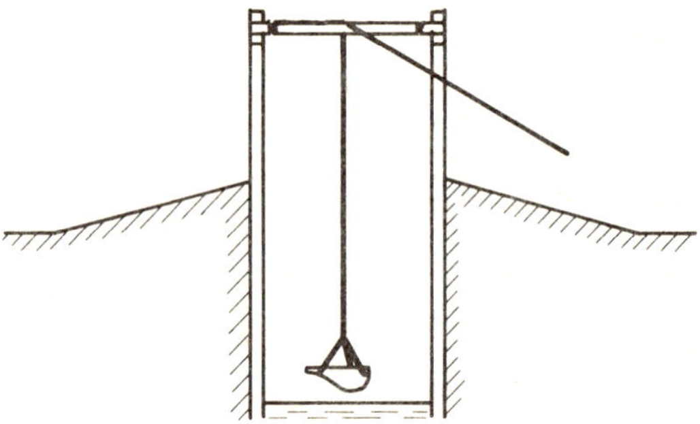

Fig. 31 Calabash on Roller

Bucket on windlass
For centuries this device has been in use and it is still being used in many places today. A rope holding a bucket is attached to a wooden roller on an iron axle with a handle. When the handle is turned the bucket goes either up or down in the well.

The bucket can be made out of a vehicle's inner tube which can be glued together at the bottom and sewn onto a wire at the top.

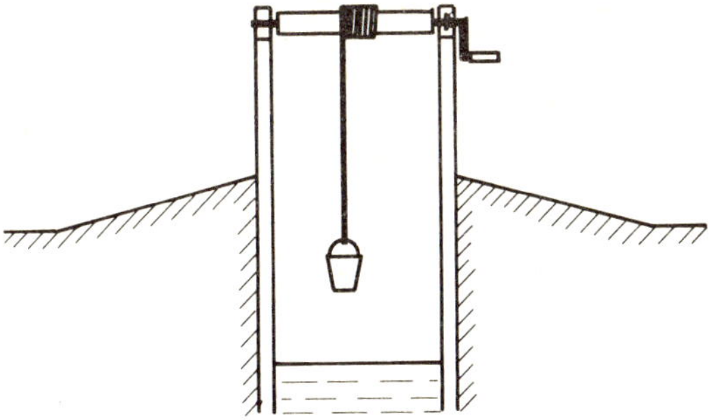

Fig. 32 Bucket on Windlass

Bucket lift in closed well
A closed well gives good protection against contamination. It works on the same principle as the lift mentioned above, the only difference is that a hosepipe is attached to the bottom of the bucket. By lifting the bucket, water will run out of the watertap on the well head.

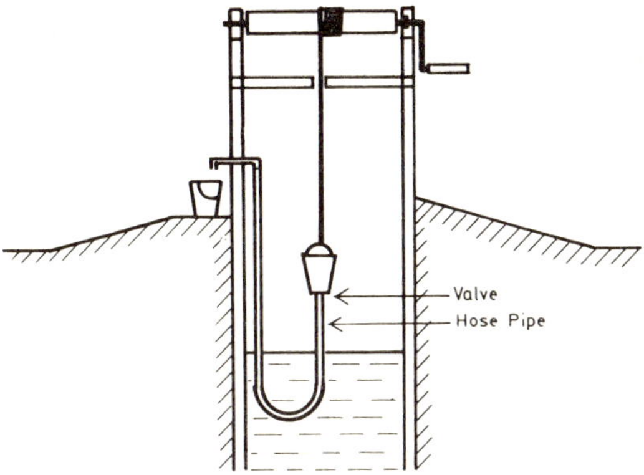

Fig. 33 Bucket Lift in Closed Well

Rope washer lift in closed well
This lift is also good for keeping a well clean. A rope with washers attached is raised through a pipe and water is thus lifted to the surface. The turning wheel can be made out of two half tyres bolted together.

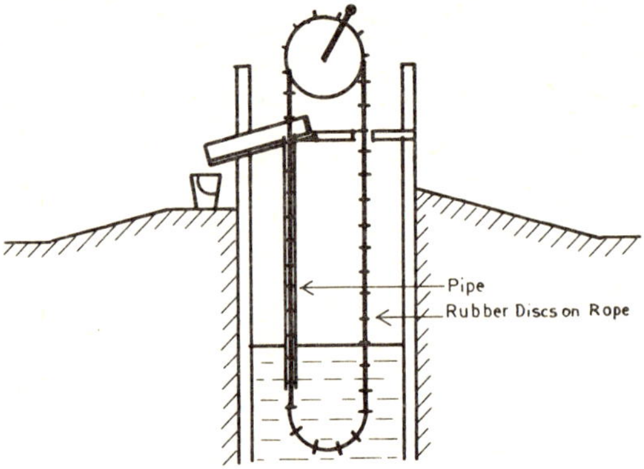

Fig. 34 Rope Washer Lift in Closed Well

Egyptian shaduf
This has been in use since the early days of the Pharaohs.

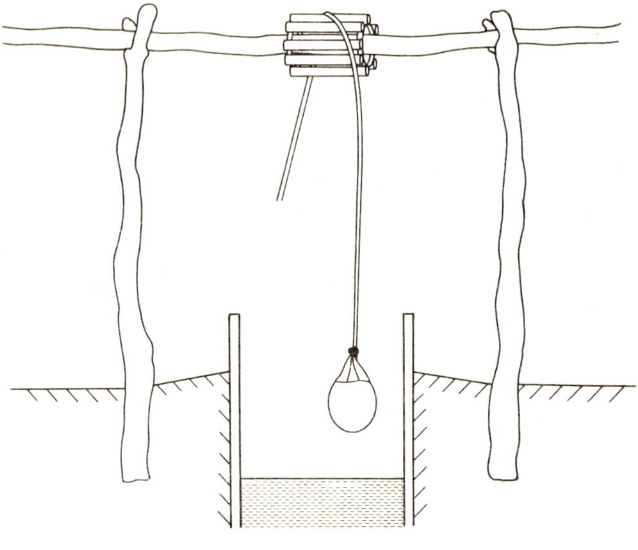

Fig. 35 Egyptian Shaduf

6 Choice of Design

Summary of possibilities

We have now been through four types of water supply system which function without pumps and pipes, and it may be helpful to summarise them:

a) Catchment of roof run-off into tanks for domestic usage.
b) Catchment of compound and road run-offs into a half sphere tank or waterholes for water for livestock, biogas, etc.
c) Catchment of field, valley and riverbend run-offs into surface and subsurface reservoirs for irrigation or construction of dams and wells.
d) Construction of wells and simple lifts for water for all purposes.

Choosing Solutions

Now your question might be: 'Which system should I start with?' and the answer will be: 'Select easy-to-build and cheap-to-finance solutions, because they will give you water and experience for a minimum of effort'.

For example build these three systems:

a) a granary basket tank for household water,
b) divert run-off from your road into a clay-covered waterhole for your livestock and fields,
c) sink an oil barrel well for all purpose water.

By doing so your investment of cash and time will not ruin your budget and, if you happen to make a beginner's mistake, it will be simple to rectify.

Simple solutions cost little money

Water, and not time, is worth a lot in arid lands. Therefore it is wise to select labour-intensive solutions which, as soon as possible, can produce a surplus of water and crops for sale.

For financing the three suggested water systems you need to sell three goats only. The benefit of having water near your house and a surplus for sale will rapidly compensate you for having three goats less.

After a rainy season or two you might start thinking of expanding your rain catchment systems to include, for example a dam, but then you must have permission from the authorities concerned before you start such an enterprise.

Procedure of dam construction

Before you embark on construction of any dam, be it a surface or a subsurface reservoir, you must consult officials of your District Water Authority. They will probably ask you to fill in some forms with data concerning your proposed dam construction. There may be a small fee to pay at this stage. The data will then be submitted to the Ministry concerned who, after approval, will make known your legal right to construct a dam on your land.

In some cases, Governments will even subsidise some of the expense of your dam construction.

Limitation of evaporation

In a hot climate the evaporation rate can exceed 1 cm per day. This means that a water depth of 2 metres can evaporate away completely within 7 months, if no measures are taken to prevent this.

It might be tempting to plant some shady trees around a water tank, but do not to this. The roots of a tree will go in searching water under the foundation of the tank and break it up.

Plant trees at a certain distance from a dam reservoir only for reducing the wind, because if planted too near to the water the trees will transpire more water than will evaporate from the surface. No appropriate method of cooling down a dam surface has yet been invented. Attempts have been made to plant water lilies, duck weed and algae, but they were all eaten by fish or just died out.

More successful results have been achieved by roofing cylindrical and half sphere tanks with chicken wire and letting passion fruit and loofah plants cover the wire with their green leaves. At the moment, a cover of old sacks sewn together and tied onto the chicken wire is being tested, which seems to be functioning even better.

A tank can, of course, be roofed with corrugated iron sheets, but this will raise the cost of the tank. Another solution is to build such a tank, not outside but inside the house. In addition to solving the problem of roofing and security of the water, it also regulates the climate in the house, making the days cool and the nights warm.

Prevention of contamination

Storage of drinking water

Long term storage of drinking water does not give rise to problems as long as some measures are followed.

a) Tanks should always be properly cleaned and washed before the start of the rains.
b) To prevent lizards and other small animals from drowning in a tank, a fine wiremesh should cover the top of the tank. The wire should be so fine that it excludes mosquitoes from breeding in the water.
c) Chemicals might be added to the water, but many people prefer to keep a few tilapia fish in the tanks. They will eat all insects and algae.

Surface water

Water in a surface dam or in a waterhole always attracts people and animals. Children play in the water, women wash clothes and fetch drinking water, men bath, and domestic and wild animals drink, bath and relieve themselves in the same water.

Let us imagine that out of all these people and animals not a single one is carrying infectious bacteria into the water. Even then, there is still a health hazard, because mosquitoes will breed and transfer malaria, snails will transmit bilharzia and several other diseases which can easily be transmitted to persons who have been in contact with that water.

The picture could be different.

Keep people and animals out of the water by fencing it with cactus and thorny bushes. Dig a ditch leading from the water, underground and beneath the fence, to three small waterholes outside the fence. The first hole for people's drinking water, the second for animals' drinking water and the third hole for children's play, women washing clothes and people bathing.

Keep the water and the area free of mosquitoes and malaria by raising tilapia fish in the water. The bilharzia snails will probably be eaten by ducks, which are also very effective eradicators of snakes, because they eat frogs, the snakes' main diet, and the snakes' young.

Another way of controlling the misuse of a surface reservoir is to sink a shallow well near the reservoir and draw all water from there.

Underground water from wells

Health conditions at a well site are controllable, especially if a man, selected by the community, is in charge of the well.

Water from a well is normally clean and it will stay clean as long as the following measures are observed:

a) The area around the well should be elevated, so that surface and waste water will not flow into the well.
b) To prevent seepage from polluted water finding its way into the well, all use of water **must** take place downstream and at a distance from the well.
c) Nobody should be allowed to enter the well or to throw anything into the well.
d) To avoid dirty containers contaminating the well, only one bucket roped onto a windlass should be used for drawing water.

If the well, in spite of precautions, gets contaminated it should be disinfected with chemicals fit for that purpose.

Reduced use of water

Many people think that there is no end to water, while other people have experienced many shortages. In all cases we should preserve what we have got, for nobody knows when a shortage will hit us.

Water for cooking can be reduced by cooking in pressure cookers, hayboxes or just by keeping a close-fitting lid on a cooking pot.

Bathing water can be minimised by using a shower. Hang a tin in the bathroom ceiling, make a few nail holes in the bottom of the tin, pour water into the tin, and your shower is ready.

A water closet consumes about half of all domestic water, if it is not adjusted to work on less volume. Bend the arm holding the floating ball downwards and consumption will be greatly reduced.

Re-use of water

Water should be re-used several times at your homestead before you give it back to its natural cycle.

Water left over from cooking is rich in nutrients. If you do not use it yourself, then cool it down and give it to your poultry. They will reward you with healthy eggs and meat.

Used bathing water containing little soap can also be used for poultry. Soapy water from bathing and washing is useful for small-scale irrigation of bananas and sugar-cane around your house.

Flush-out from a water closet can either enter an earth septic tank from where it can irrigate by seepage, particularly fruit trees, or it can enter a biogas digester. After having passed through a biogas digester, it can either be used as fertilizer for an orchard or as fish feed.

It may not sound appetising, but the Chinese people have practised bio-dynamic farming methods, even without biogas digesters, for thousands of years, so it is safe and properly tested.

7 Build-It-Yourself Constructions

The last chapter of this manual shows in step-by-step drawings and instructions ways in which you can solve your water problems. To assist you in estimating the cost of the constructions in this section, there are specified lists of working time and materials.

To obtain a good result from the construction work, it is important to follow the instructions and not to use less cement or wire reinforcement than described.

Construction of a 23,000 litre roof catchment tank for domestic water

Working time
32 working days for 1 man

Materials (including making the blocks)
31 bags of cement
60 wheelbarrows (7½ tonnes) of coarse sand
30 wheelbarrows (4 tonnes) of 1 cm ballast (small, hard stones)
20 wheelbarrows (2½ tonnes) of 4–5 cm ballast
270 m of barbed wire
2½ m of any straight water pipe
1 m of any water pipe with a water tap
½ m guttering
3½ kg of 2″ nails
1 kg of 2½″ nails
5 kg roofing nails
5 kg water proof cement
4 lengths, each 5 m long, of 8 × 5 cm (3″ × 2″) timber
1 length, 3½ m long, of 13 × 2.5 cm (5″ × 1″) timber

16 sheets of galvanised corrugated iron sheets
5 litres of white oilpaint
½ litre of red oilpaint
1 litre of used motor oil
13 drums of water

1 Excavation of foundation

Working time
1 man in 1 day.

Materials
2½ m of any straight water pipe.

Instructions

a) Select a site for the tank which has no tree roots or filled-up holes. The site should be next to the house and preferably at the lowest corner of the roof.

b) Hammer a wooden peg into the ground at the centre of the site, which is 2.5 m from the wall of the house.

Attach a string to the peg. Tie a loose peg to the other end of the string at 2.25 m from the centre of the fixed peg.

Draw a circle on the soil. This is the outline of the foundation.

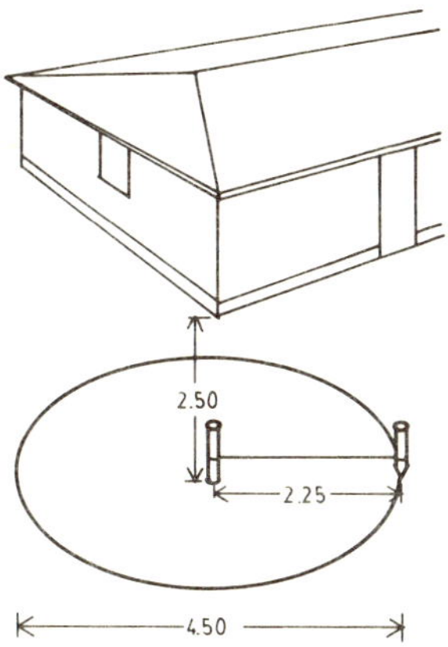

Fig. 36

c) Dig out all mulch and grass roots within the circle but don't remove the fixed peg. Excavate the foundation floor until there is 2.3 m between the lowest point of the gutters and the floor, which must be exactly horizontal. Check the level with a long spirit-level placed on a straight board of timber.

d) Remove the fixed peg and place a straight water pipe in its position. Hammer the pipe about 25 cm into the ground and make sure that it is exactly vertical. Check it with the spirit-level.

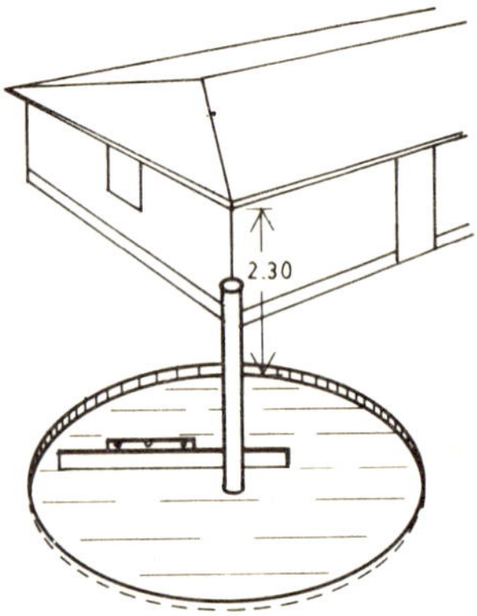

Fig. 37

2 Concreting the foundation

Important: This concrete work must be completed within *one* day.

Working time
2 men in 1 day.

Materials
6 bags of cement.
12 wheelbarrows of coarse sand.
20 wheelbarrows of 4–5 cm ballast.
160 m of barbed wire.
1 drum of water.

Instructions

a) Mix 3 bags of cement with 6 wheelbarrows of sand until it is a uniform colour. Then make a depression in the middle of the pile and pour water into it. Mix in water until the whole pile has reached a consistency similar to that of thick porridge. Then add 10 wheelbarrows of ballast and mix it thoroughly so that all the ballast is coated with cement. Pour the concrete into the foundation and level it out in an equal layer.

b) Lay upon the concrete lines of barbed wires spaced 20 cm apart. Then place at right angles a second layer of barbed wires, also spaced 20 cm apart.

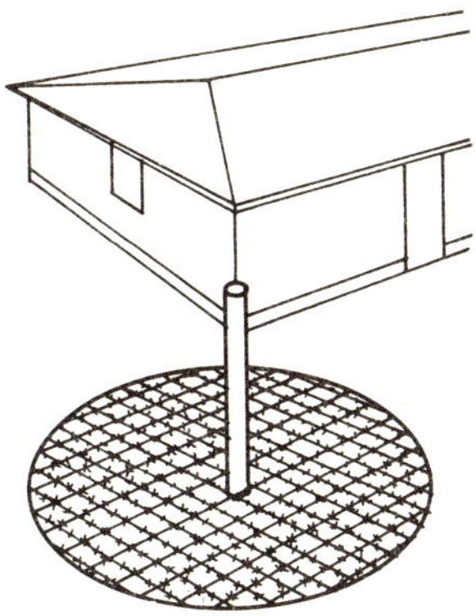

Fig. 38

c) Mix a second portion of concrete similar to the one described above. Pour the well-mixed concrete onto the foundation. Level and compact it. Use a spirit-level and a straight piece of timber to make it exactly level. Roughen the surface with a broom. Check that the pipe is still in an exact vertical position. Cover the concrete work with grass or wet sacks and keep it moist and under shade for a couple of weeks.

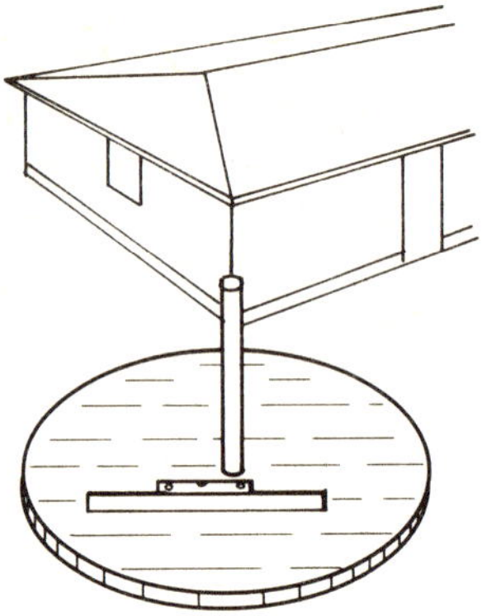

Fig. 39

3 Building the first course of the wall

Working time
2 men in 1 day.

Materials
27 blocks, 13 × 23 × 46 cm (See fabrication of blocks on page 55.)
½ bag of cement.
1 wheelbarrow of sand.
4 buckets of water.
14 m of barbed wire.
½ kg of 2″ nails.
1 m water pipe with a water tap. Any dimension can be used.

Instructions

a) Lay out 27 blocks in a circle 10 cm from the outline of the foundation and space them equally apart. Each block must be situated so that the side where the lid was placed during its making will be at the outer side of the tank wall.
b) Wrap a small ring of wire around the pipe in the middle of the foundation and tie a piece of string to the ring. Make a knot on the string exactly 2.15 cm from the centre of the pipe. This knot shows the correct position of the outer corners of each block in the tank wall.

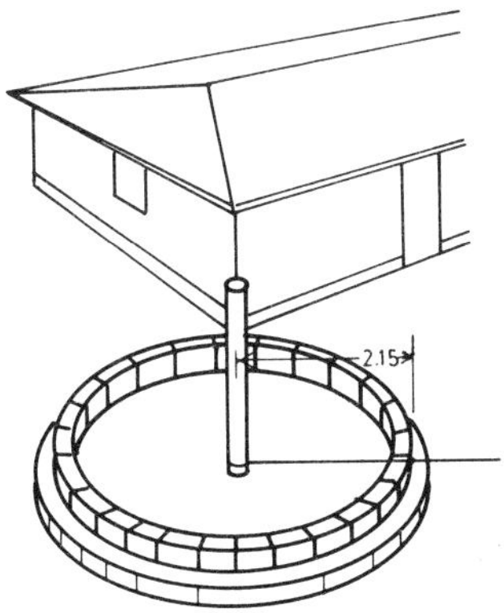

Fig. 40

c) When all blocks are laid out, then water the blocks lightly. Mix ½ bag of cement with 1 wheelbarrow of sand.
   Lift one block at a time and place a layer of mortar in its place. Lay back the block and bring it into a level position by using a spirit-level and a hammer. Use a piece of timber on the block to protect it from the hammer.

d) When the block has been cross-checked for level in both directions, then use the knot on the string to determine the exact positions of the outer corners of the block.
   Follow this procedure for all 27 blocks. Remember to check that the distances between the blocks are the same and that the pipe in the centre of the foundation is still in an exactly vertical position.

e) Take the outlet pipe and screw the water tap tightly onto one end and, to avoid the pipe being blocked with mortar, put a piece of cloth into the other end.
   Select a convenient space between two blocks and pour a little mortar between them. Press the pipe as close as possible to the foundation. Fill the space between the two blocks with mortar, compact it and make it flush with the two blocks.
   Fill all other spaces between the blocks with mortar, compact it and make it flush with the blocks.

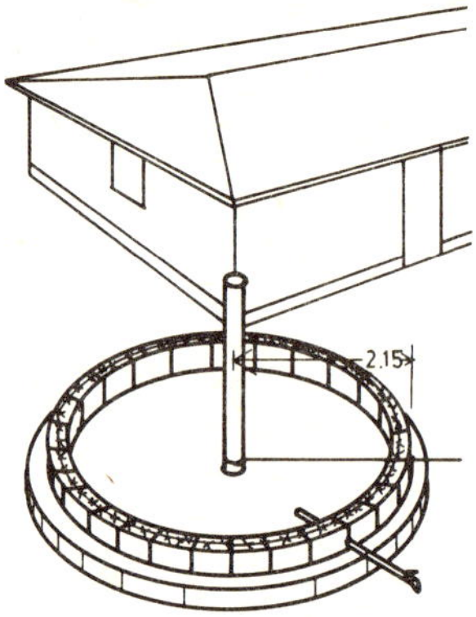

Fig. 41

f) Lay barbed wire around on the top of the blocks. Nail the wire onto the middle of each joint between the blocks. Keep the wire tight and let the two ends overlap each other by at least 20 cm.

4 Building the remaining 6 courses of the wall

Working time
2 men in 3 days.

Materials
162 blocks, 13 × 23 × 46 cm.
3½ bags of cement.
7 wheelbarrows of sand.
90 m of barbed wire.
3 kg of 2″ nails.
1 drum of water.

Instructions

a) Sprinkle the completed course lightly with water. Lay upon that course a round of 27 blocks, each one to be placed with its middle exactly over the joint of the underlying course and with the right side, the lid side, facing outwards. Sprinkle the blocks lightly again. Mix mortar and build the course as the first one was built.

b) Nail also a ring of barbed wire onto this second course.
c) Finally, tie the timbers onto the barbed wire which is lying on the top of the last course with other lengths of wire.
Each course must have a ring of barbed wire nailed onto it.

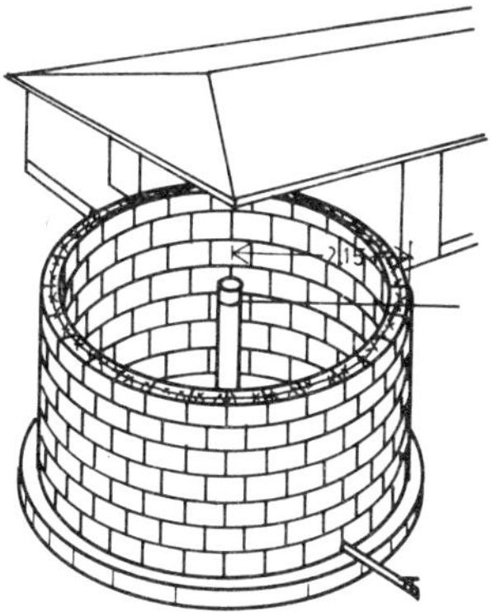

Fig. 42

5 Fixing overflow and timber for the roof

Working time
1 man in 1 day.

Materials
½ m of guttering.
4 lengths, 5.0 m long, of 8 × 5 cm (3″ × 2″) timber.
4 m of wire.

Instructions
Overflow The overflow gutter should be fixed vertically above the outlet water pipe, so that overflowing water will splash onto the concrete section around the water tap (see next page) and not erode the foundation of the bank.

Chisel out for the gutter in the top course vertically above the outlet pipe. Don't remove the barbed wire tied there but use it, together with a few nails, to keep the gutter in position.

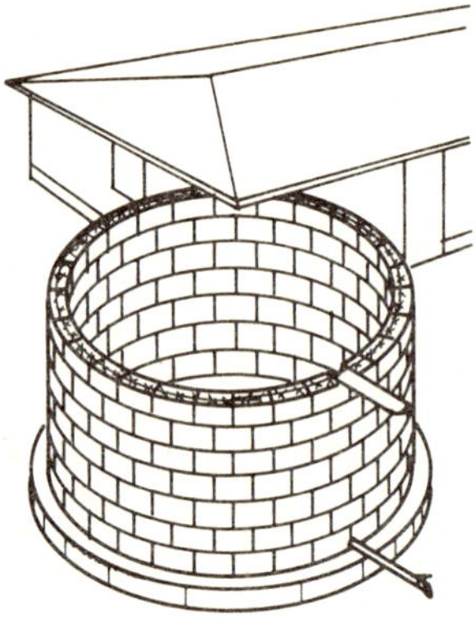

Fig. 43

Timber for the roof The purpose of the roof is to minimise evaporation which otherwise can empty a full tank within half a year. To collect run-off from the roof it should slope towards the centre of the tank.

To enable a person to enter and clean the tank, the two roof sections should be at least 40 cm apart.

a) Place the two strongest lengths of timber upon the tank. Situate them along a diameter parallel to the roof of the house and 80 cm apart.
b) In order to lift the timber from the top of the tank 4 short pieces of 5 cm timber should be placed on top of the blocks. Then place the two other lengths of timber parallel with the first two lengths but 1m apart. Raise the last two lengths by placing 4 pieces of 8 cm timber under them.

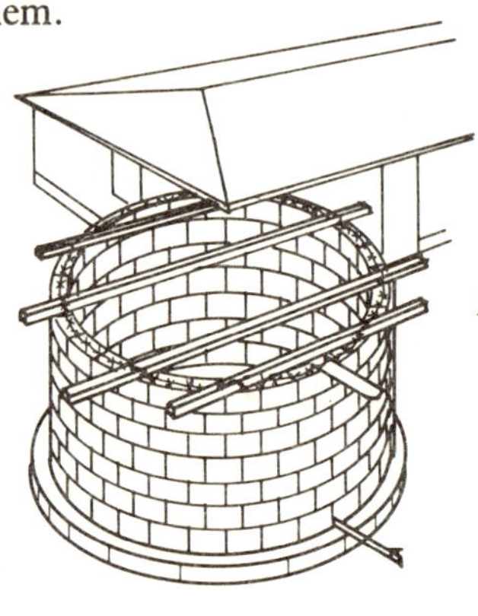

Fig. 44

c) Finally, tie the timbers onto the barbed wire which is lying on the top of the last course with other lengths of wire.

6 Roofing and water tapping section

Working time
1 man in 2 days.

Materials
16 galvanised corrugated iron sheets.
5 kg of galvanised roofing nails.
1 bag of cement.
2 wheelbarrows of sand.
4 buckets of water.

Instructions
Roofing

a) Nail the corrugated iron sheets onto the timbers on the tank. Remember to leave a space between the two roof sections big enough to allow a person and a ladder to enter the tank.
b) Cut off the corners of the roof as shown. By using the cut-off corners for the roof, it is possible to save 2 iron sheets.

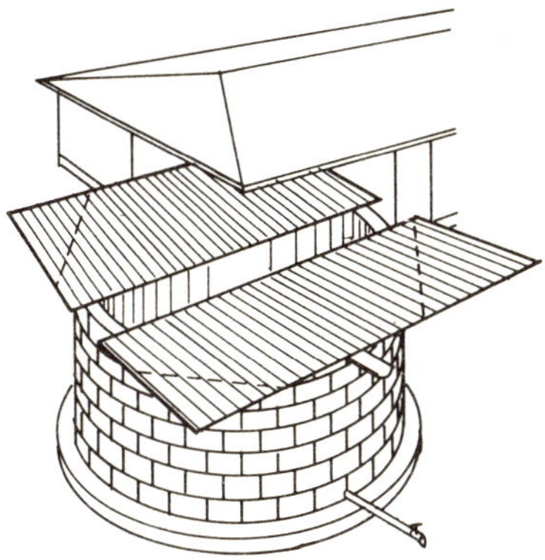

Fig. 45

Water Tapping section In order to avoid erosion around the water tap and the overflow, the place should be concreted and a ditch dug for leading away any surplus water. Under and around the water tap there must be sufficient room for a bucket and a jerrycan.

a) Dig out a square hole under and next to the water tap. Measure the size of the hole to be sure that a bucket and a jerrycan will fit easily.
b) Plaster the hole and make a small apron around it. Make a groove in the lowest corner of the apron for leading away any surplus water.

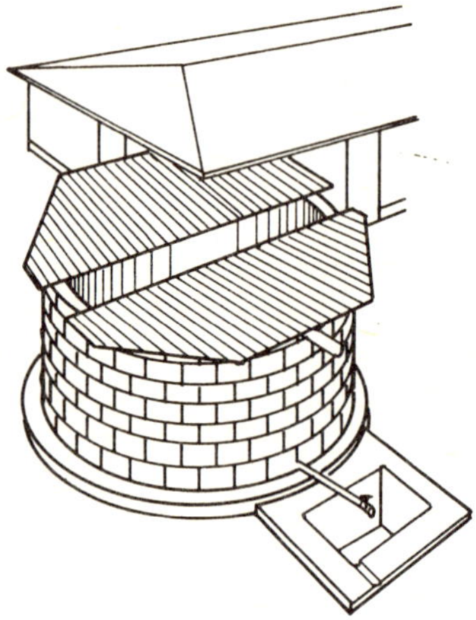

Fig. 46

c) Dig a ditch for leading away surplus water further away from the apron.

7 Interior plastering of the tank

Important. Plastering and curing must be carried out in shade.

Working time
2 men in 3 days.

Materials

8 bags of cement.
5 kg of waterproof cement.
15 wheelbarrows of sand.
4 drums of water.

Instructions

Remove the pipe in the centre of the foundation.

a) Clean the interior of the tank with a hard brush. After all the sand and dirt have been removed, dampen the tank wall throughout.

b) Mortar should be mixed in portions consisting of 1 bag of cement and 2 wheelbarrows of sand and a suitable volume of water.

   Throw well-mixed mortar onto the inside and the top edge of the tank wall with a trowel. For vertical smoothing move a straight length of timber up and down. For a finer smoothing use a float. Continue to throw on and smooth portions of mortar until a smooth surface without holes or bulges is reached.

   This work must be completed in one day.

c) Clean and dampen lightly the foundation floor. Lay upon the floor a layer of mortar, all of it sloping down towards the open end of the outlet pipe. Smooth it well, and make sure that the outlet pipe doesn't get blocked with mortar.

   When the floor is dry enough to walk upon, place a line of mortar along the corner of the wall and the floor.

   Take an empty bottle and use it as a float for making a round and smooth junction.

   This work must also be completed in one day.

d) Mix 1 kg of waterproof cement with 3 kg of cement and add water. With a steel trowel coat the entire interior of the tank and the top edge with this waterproof material. Start at the top of the tank and finish off at the floor.

e) For the next 2 to 3 weeks the tank and its plaster must be kept moist and under shade.

8 Exterior pointing of the tank

Working time

1 man in 2 days.

Materials
2 bags of cement.
2 wheelbarrows of sand.
4 buckets of water.

Instructions

a) Mix 1 bucket of cement with 1 bucket of fine sand and add water.
b) Starting at the top of the tank, plaster a section of joints. Smooth the plaster with a float and a trowel. Then, using a spirit-level, draw upon the wet plaster vertical and horizontal lines, 3 cm apart, showing the joints lying underneath the plaster.

   After that has been done remove all the plaster which is outside the lines with a trowel and neat, watertight pointing will be achieved.
c) Fill the joint between the wall and the foundation on the outside of the tank with a sloping apron of concrete.

9 Interior painting and calibration

Working time
1 man in 2 days.

Materials
5 litres of white oil paint.
½ litre of red oil paint.

Instructions

a) After the tank is completely cured and dried, paint the entire interior of the tank with white oilpaint.
b) On a visible part of the inner wall, mark with a pencil a straight vertical line.
c) Starting from the bottom of the tank, measure out every 8 cm along the vertical line. Every 8 cm mark is equal to a volume of 1,000 litres of water. Mark the lowest point with "1,000 L" and the next lowest with "2,000 L" and so on until the top is reached.
d) Paint every 1,000 litre mark and the figures with red oil paint.

Fabrication of concrete blocks

Construction of the tank described requires 189 blocks size 13 × 23 × 46 cm (5″ × 9″ × 18″). Each one of the blocks must be of strong and well-cured quality.

Quarry blocks can be used but clay bricks are not recommended.

Working time
1 man in 8 days.

Materials
3.5m of 13 × 2.5 cm (5″ × 1″) timber planed on 4 sides.
1 kg of 2½″ nails.
1 litre of used motor oil.
10 bags of cement.
20 wheelbarrows of sand.
30 wheelbarrows of 1 cm ballast.
5 drums of water.

Instructions

a) Cut the timber into pieces as shown, making sure you cut at right angles. Nail the timbers together and reinforce the form with iron straps or wires.

Coat the form with oil.

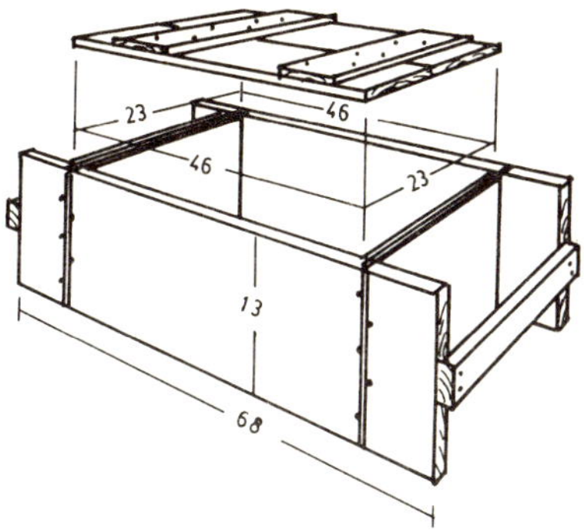

Fig. 47

b) Find a suitable place, preferably under a shady tree, for mixing concrete and forming blocks.

Clean and level the surface of the soil and spread out a thin layer of sand.

c) Mix 1 bag of cement with 2 wheelbarrows of sand. Mix in as little water as possible and then add 3 wheelbarrows of ballast. The final mixture should be of a consistency which only allows a few drops of water to be squeezed out of a handful of concrete.

d) Place the form on the soil and fill it with concrete. Compact and level it off flush with the top of the form.

e) Place the lid upon the levelled concrete without touching the timber of the form.

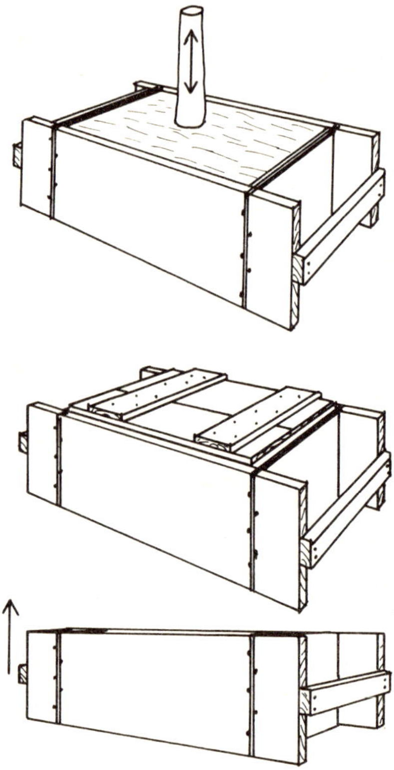

Fig. 48

f) Place one foot on the lid and your hands on the handles at the ends of the form. Now lift the form while still keeping your foot on the lid.

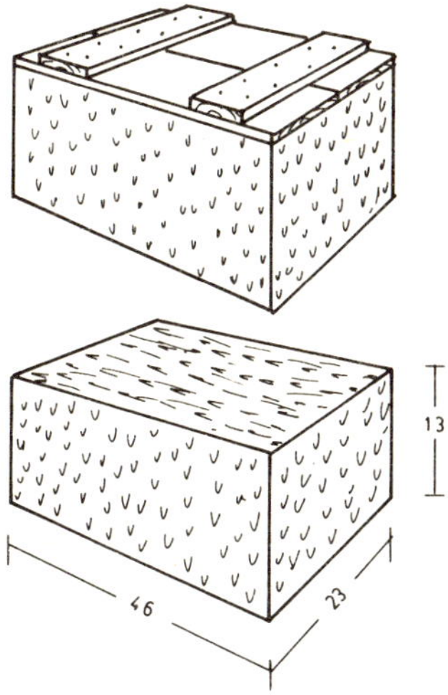

Fig. 49

g) A concrete block will emerge. Cover it with grass or paper.
Keep this block and the following 188 blocks moist and in the shade for a couple of weeks.

Construction of a 51,000 litre compound catchment tank for livestock water and fish farming

Working time
2 men in 12 days.

Materials
25 bags of cement.
120 sq m of 1″ chicken wire.
1 length, 4 m long, of any straight galvanised pipe.
8 lengths, 0.8 m long, of any iron pipe or iron rod.
50 m of 3–5 mm galvanised wire.
3 kg of waterproof cement

20 kg of bitumen.
88 wheelbarrows (8 tonnes) of coarse sand.
5 drums of water.
1 roll of barbed wire, 12½ gauge.

Instructions

1 Excavation

a) In order to get the tank filled with rainwater run-off from your compound or road, the tank should be situated at the lowest point of the catchment area.
The site has to be free of trees and roots.

b) Fix a wooden peg in the centre of the selected site and tie a string to it. Tie a loose peg to the string 3 m from the centre of the fixed peg. Draw a circle on the soil which will indicate the outline of the tank.

c) If you are satisfied with the position of the tank, then draw a second circle 1 m from the centre of the fixed peg.

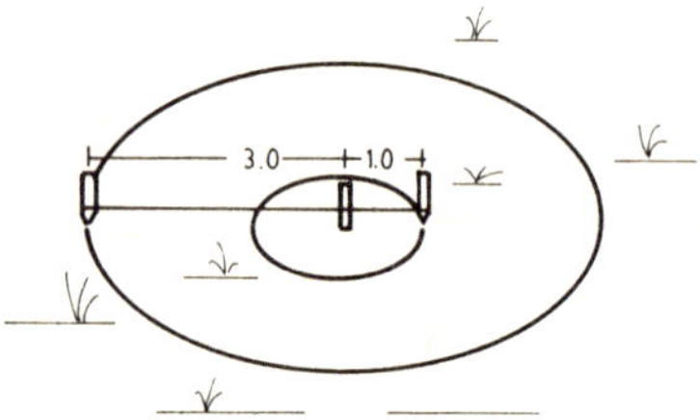

Fig. 50

d) Dig all soil out within the 1 m circle to a depth of 3 m.

e) In the centre of the bottom of the hole, dig a small hole 30 cm deep and about 30 cm wide.

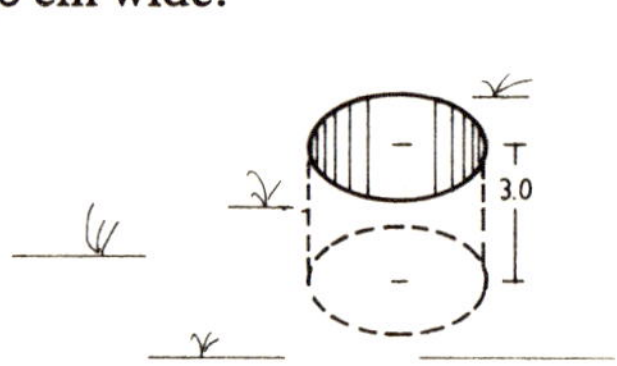

Fig. 51

f) Pour 10 cm of concrete into that small hole, place a straight 4 m length of any galvanised iron pipe on the concrete and fill concrete all around the pipe until the small hole is filled up.

Position the pipe exactly vertically and check that it stays that way during the whole construction period.

After completion of the tank that pipe will be standing there as a pillar for the wire roof.

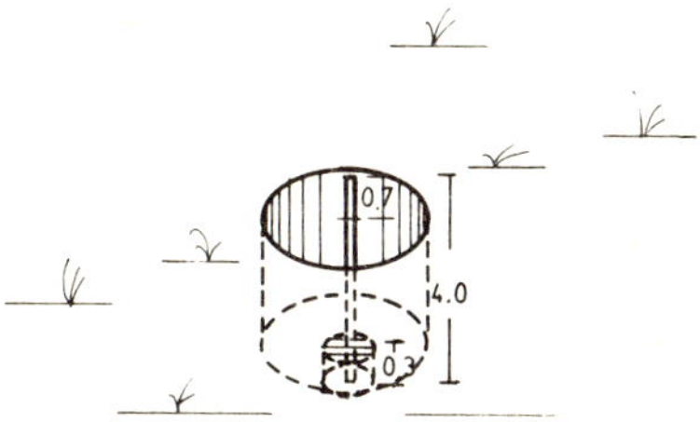

Fig. 52

g) In case the line of the big circle has disappeared during digging, draw it once more.

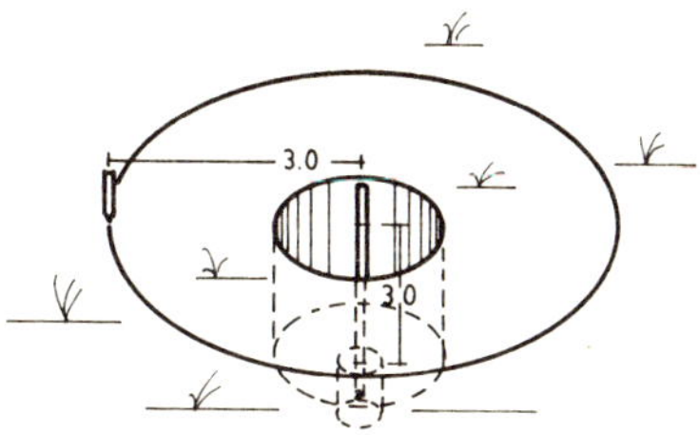

Fig. 53

h) Mark the pipe 3 m above the concrete in the bottom of the hole. This mark should also be level with the surface of the ground.
i) Cut a 3 m long measure stick for measuring the distance from the mark on the pipe to the wall of the hole to be excavated.
j) Start digging soil out along the drawn circle and continue downwards. Use the measuring stick to keep correct distance from the mark on the pipe.

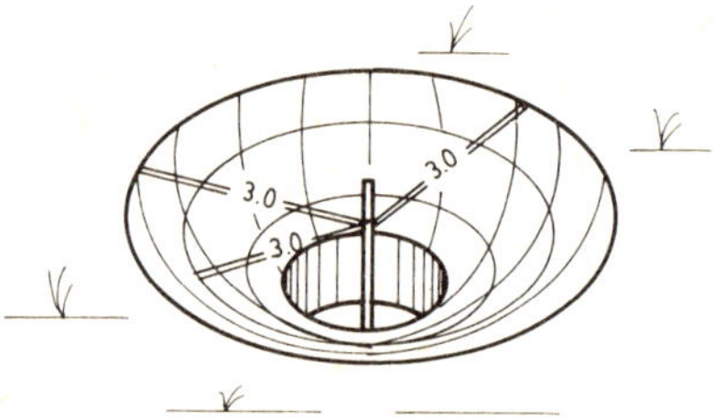

Fig. 54

k) You should now have a half-sphere excavation. Remove any bulges. Holes, if any, will later be filled with mortar.

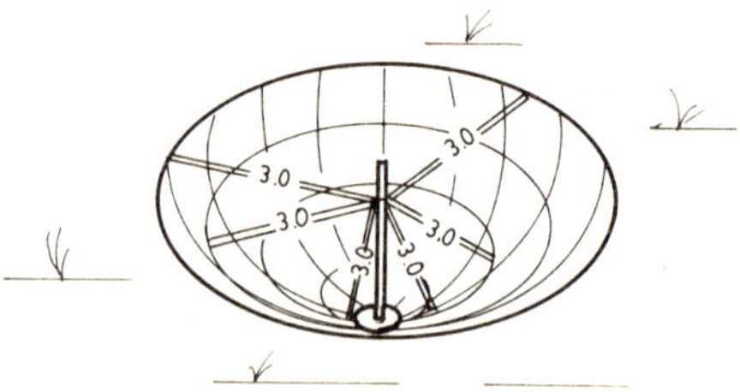

Fig. 55

2 Plastering

Important: Each of the coatings described below must be completed within *one* day.

The coatings must also be kept well shaded and moist by covering them with wet sacks when under construction and for 2 weeks after completion.

a) *1st Coat:* Mix about equal volumes of water and cement to a creamy consistency. Brush it onto the surface of the excavation and its edge so as to bind the soil particles.

b) *2nd Coat:* Mix 1 bag of cement with 2 wheelbarrows of sand and water. Throw this mortar onto the coated surface. Repeat this process until a 3–4 cm layer is produced.

c) *3rd Coat:* Starting at the upper edge of the excavation, roll out and nail the barbed wire onto the plaster. Nail two rounds at the top edge. Then continue rolling and nailing the barbed wire in a descending spiral with the strands at intervals of roughly 30 cm.

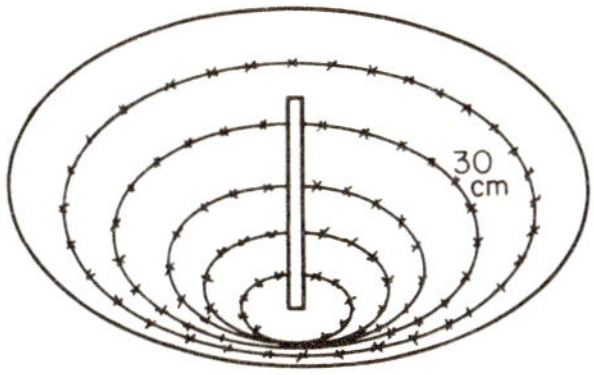

Fig. 56

d) Nail 1″ chicken wire onto the plaster which has to be totally covered with chicken wire. All overlaps must be at least 15 cm.

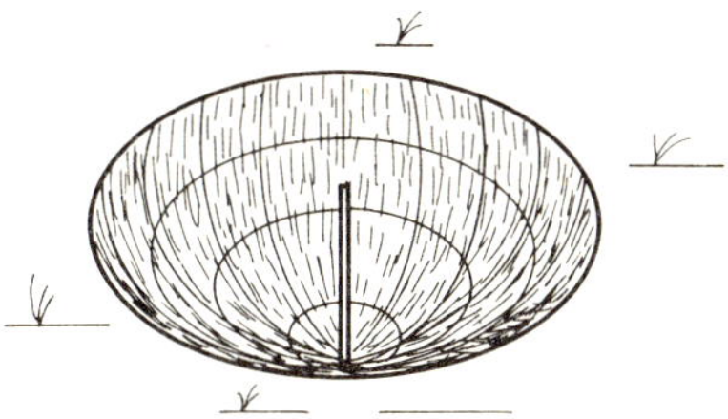

Fig. 57

e) *4th Coat:* Mix mortar as in 2nd coat and throw it onto the chicken wire in a 3–4 cm thick layer. Smooth it with a float and a brush.

f) *5th Coat:* Mix 1 kg of water proof cement with 3 kg of cement. Coat the surface with it. Repeat this process until the whole edge is coated.

Start at the top edge using a steel trowel.

g) *6th Coat:* When the concrete work is completed and dried, it should be coated twice with bitumen.

3 Roofing

a) Ram 8 pieces of 1 m long iron rod or pipe down around the edge of the tank. They should be spaced equally apart and slope away from the centre of the tank.

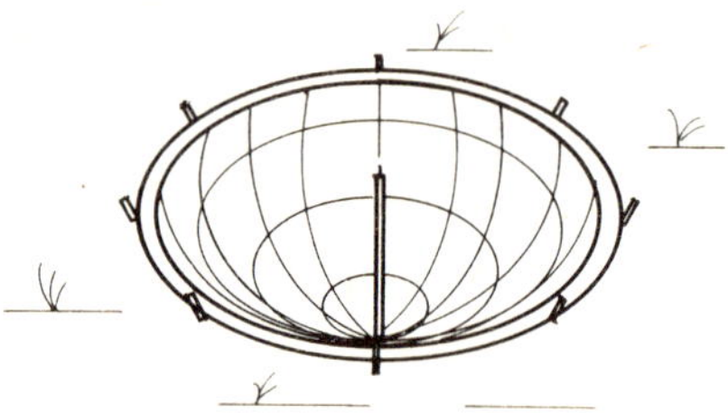

Fig. 58

b) Fill the centre pipe with mortar and insert a short piece of iron rod in the top of the pipe protruding about 5 cm.

Tie a 3–5 mm galvanised wire from that short iron rod to each of the 8 irons around the tank.

c) To ensure that no child or animal can fall into the tank, it is now essential to bind chicken wire onto the wire structure.

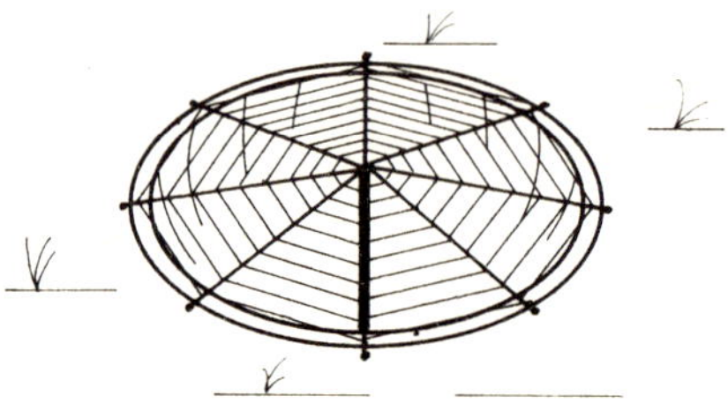

Fig. 59

d) To limit evaporation the chicken wire can be covered with sacks or green creeping plants.

Construction of a concrete gravity dam for sinking wells and seepage irrigation

Working time
2 men in 80 days.

Materials
150 bags of cement.
35 tonnes of sand.
35 tonnes of ballast.
50 tonnes of boulders (big stones).
50 drums of water.

Instructions

1 Design and approval

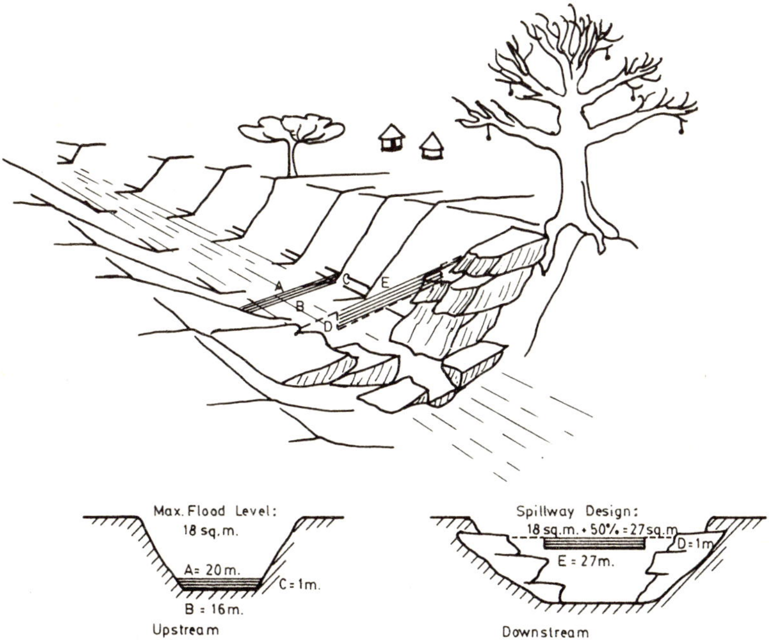

Fig. 60

a) Find a suitable site in a semi-permanent riverbed. Preferably the site should have a solid bedrock in the riverbed and hard rocks protruding from both sides. Ideally the riverbed upstream should be almost horizontal.

b) Measure the lengths (A and B) and the height (C) of the greatest area of flooding which has occurred in the riverbed. Multiply the figures and add 50% for security reasons. You will now know the size of a spillway necessary for allowing exceptionally big floods to pass over the dam wall safely.

c) Dig out all sand and loose stones covering the bedrock. Draw a level string across the rocks to indicate the upper edge of the spillway. Draw a second string 1 m below the first string to show the bottom of the spillway, which is also the top of the dam wall.

Measure the distance from the lower string to the bedrock. This measurement is the height of the dam wall and in the example shown it is 2.5 m. This figure, plus 0.3 m gives the thickness of the dam wall at the bottom; namely 2.8 m.

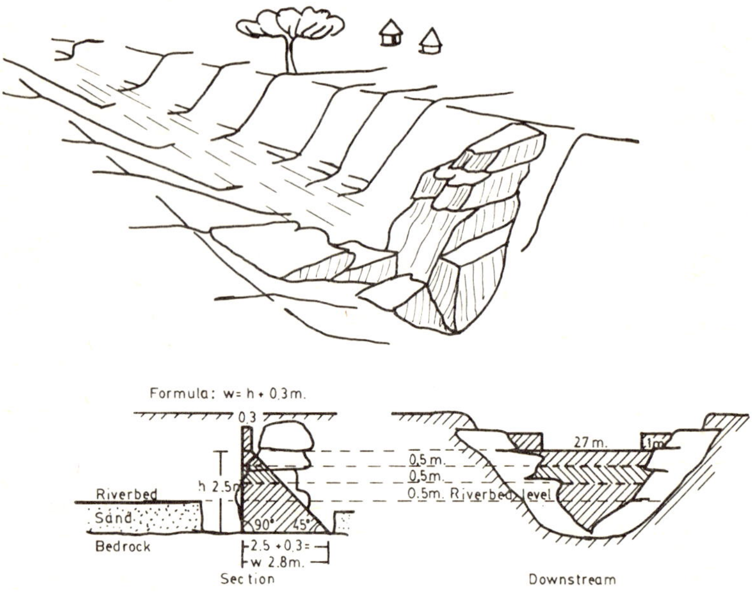

Fig. 61

d) Provided that it still seems feasible to build a dam in that place, then mark the various measurements taken and their points with paint on the rocks.

Draw a sketch of the proposed construction and ask your District Water Bailiff to assist you in getting the proposed dam construction approved.

e) In order to get the reservoir silted up with as coarse sand as possible, for the reason of holding a maximum of water, the dam wall should be built in 0.5 m sections above the riverbed. Each section should be allowed to silt up before the next section is built. Although this method takes longer, it will enable the reservoir to hold twice as much water.

2 Construction of the first section

a) When you have the approval of the Apportionment Board, you can buy the cement and start constructing the first section. Remove all bushes, trees and their roots from the construction site.

Redraw the strings which indicate the spillway. Draw a third string near the bedrock and vertically below the spillway strings.

b) Clean, sprinkle, and coat with cement slurry the rocks where you are going to work with concrete. If there are any cracks in or between the rocks, they must be blocked carefully with mortar made by mixing 1 bag of cement and 3 wheelbarrows of sand. Use the same mixture for concrete but add 3 wheelbarrows of ballast.

c) Build along the string at the bottom a line of big stones with their plane sides facing upstream. Fill all spaces between the stones and the rocks with concrete and smaller stones.

Stones and rocks should not touch one another but should have mortar or concrete in between.

d) When the wall has reached a height of 0.5 m above the riverbed, then plaster to a finish the upstream side of the wall. On the downstream side fill concrete and stones up to the backtoe of the dam wall which is 2.8 m from the upstream side. The slope of the downstream side should be 45°, which is the same as saying that for every 0.5 m height the dam wall must be 0.5 m thinner.

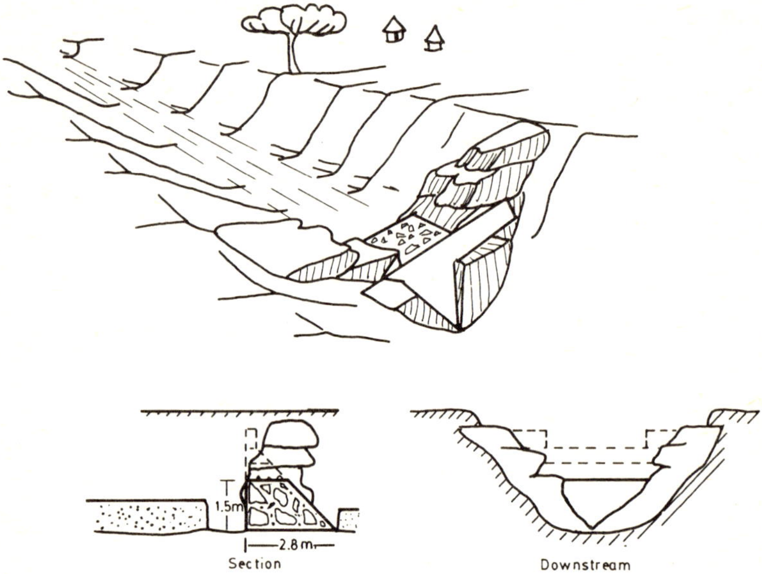

Fig. 62

e) To achieve a good binding to the next construction section you should make a rough surface at the top of the first section.

3 Construction of the second section

a) After a flood has silted up the reservoir, you can continue with the second section, which is built in the same way as the first section.
Remember to clean, sprinkle, and coat with cement slurry all the surfaces to be built onto.

b) Build the wall 0.5 m high and 0.5 m thinner at the top. Plaster the upstream side and leave a rough surface at the top of the dam wall.

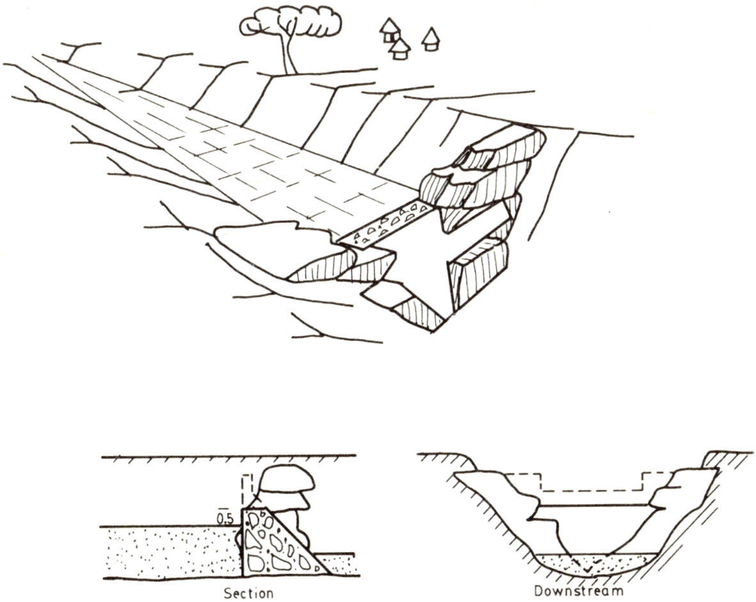

Fig. 63

4 Construction of the third and final section

a) After another flooding, the third and final section can be built as with the former two sections. The thickness at the top of the wall will be 0.3 m.
b) This 0.3 m thickness is extended 1 m above both ends of the dam wall to the spillway.
c) Plaster to finish the whole dam wall on all visible sides.

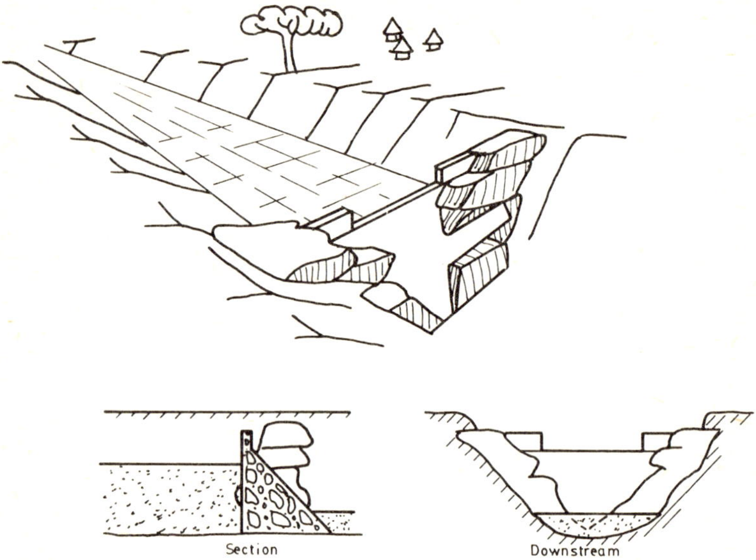

Fig. 64

5 Usefulness of the reservoir

a) Floods have now filled the reservoir with sand but water will not yet be stored there for the maximum length of time. This will be achieved after several more fillings which will saturate the adjacent land. That process will raise the subsoil watertable, which prolongs the growing season and eliminates the risk of crop failure due to drought.

b) Establishment of an orchard, vegetable garden and fodder trees for livestock and for fertilising your field is also feasible around the reservoir.

c) Due to the advantage of holding the water in a subsurface reservoir, it does not tempt children or livestock to invade the reservoir.

d) It has also the advantage of not being suitable for mosquito breeding, which minimises the danger of malaria. The reservoir also will not attract snakes because there will be no frogs living there.

e) The most hygenic method of drawing water for domestic and livestock purposes is to sink a shallow well in a riverbed near your house.

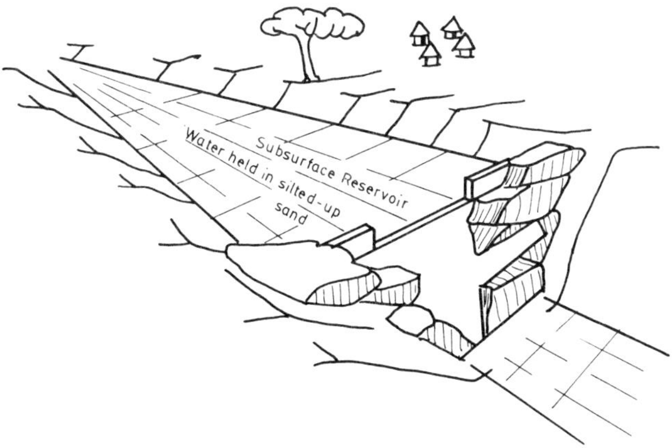

Fig. 65

Construction of a sinking well for domestic and livestock water

Working time

Depends mainly on the depth of the well and the type of soil. As a guideline it will take 2 men 30 days to construct and complete a 5 metre deep well with head and slab.

Materials

Materials for 5 metre deep well with head and slab:

17 bags of cement.
30 wheelbarrows (3½ tonnes) of sand.
10 m of 12 mm iron rod.
150 m of 5 mm iron wire.
1 form for blockmaking (see page 77).
1 windlass or another lifting device.
5 drums of water.

Instructions

1 Siting and preliminary excavation

a) Having found a potential site for a well, mark out a square measuring 140 cm × 140 cm and dig as far down as possible.

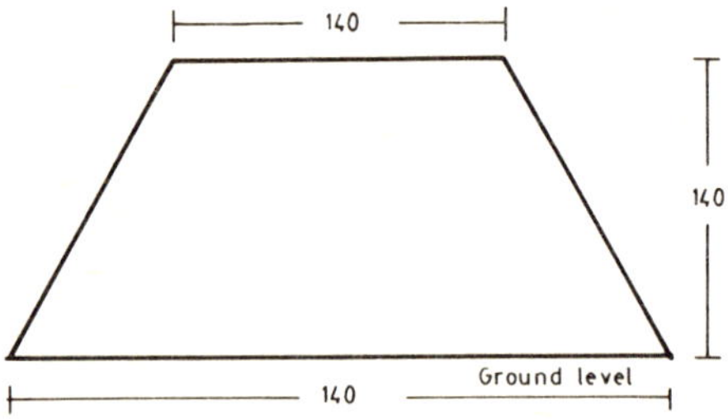

Fig. 66

In case the soil is sandy or otherwise unstable, it is necessary to widen the excavation at the top to eliminate the risk of the walls collapsing in on you.

b) If you haven't found any evidence of water in the bottom of the excavation, you have to find out whether there is water further down in the soil or look for a better site.

To find out about that you can do three things: 1) widen and deepen the excavation; 2) hammer down a length of 12 mm iron rod in several places; or 3) drill a few holes with a hand auger.

2 Concreting the foundation ring

a) When you are sure that a well can be sunk further down into a waterbearing layer (an aquifer), the work can proceed.

b) Nail together pieces of timber as shown and use them to excavate a round and level groove in the bottom of the excavation.

This groove will be the form for concreting a reinforced cutting ring which will also act as a sinking foundation for the well.

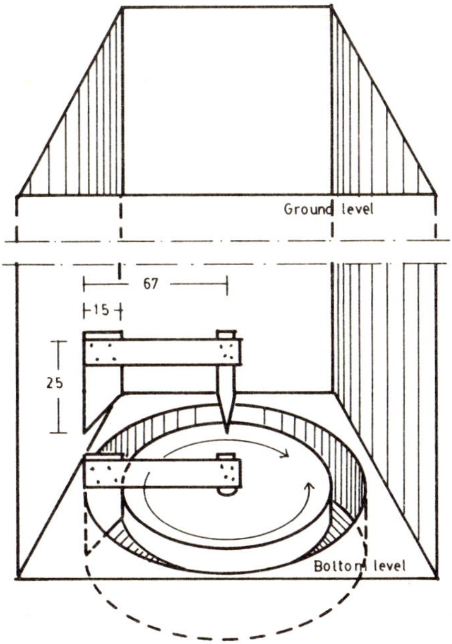

Fig. 67

c) Mix ½ bag of cement, ¾ wheelbarrow of sand, ¾ wheelbarrow of ballast and some water.

Pour the concrete into the toe of the groove. Lay an iron wire, tied together as a ring, into the concrete. Fill up half the groove with concrete. Take another iron ring and tie 16 lengths of wire, each 50 cm long and with their ends bent, onto it and lay it into the concrete with the 16 wires spaced equally apart.

Fill the groove with concrete and press another 2 iron rings into the wet concrete. Smooth the concrete and let it harden for about a week.

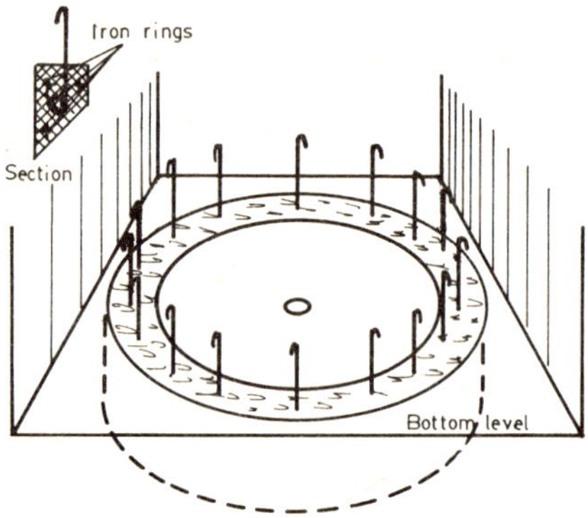

Fig. 68

3 Building the well up to ground level

a) Clean the surface of the concrete ring. Mix mortar consisting of ½ bag of cement, 1 wheelbarrow of sand, and water. Build the first round of blocks upon the concrete ring. Each block has to be flush with the inner side of the concrete ring and each of 16 wires must be placed in the middle of the holes in the blocks.

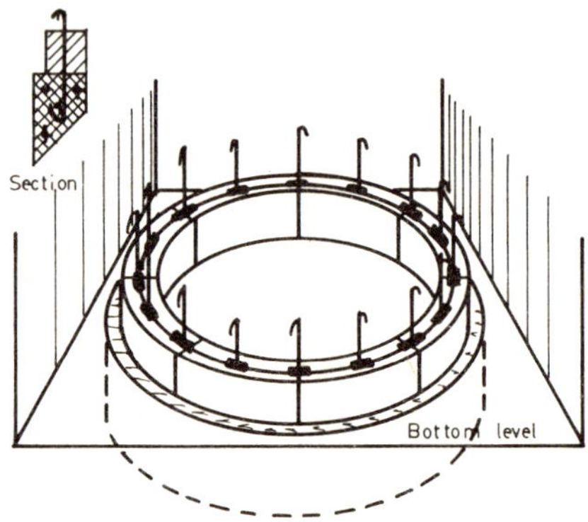

Fig. 69

When all blocks and wires are situated correctly, then fill the holes in the blocks with mortar. Compact and level the mortar.

b) Take another 16 lengths of iron wires, each 1 m long, and bend their ends. Fasten each of these wires to the 16 wires in the concrete ring.

Place the second and third courses of blocks loosely on the first course and with the iron wires going through all the holes. Don't cement the blocks together.

c) At the third course a step must be built in. Cut a 70 cm long piece of 12 mm iron rod and bend it as shown. Chisel holes for it in the blocks and fasten it with mortar using the blocks above it to keep it embedded solidly.

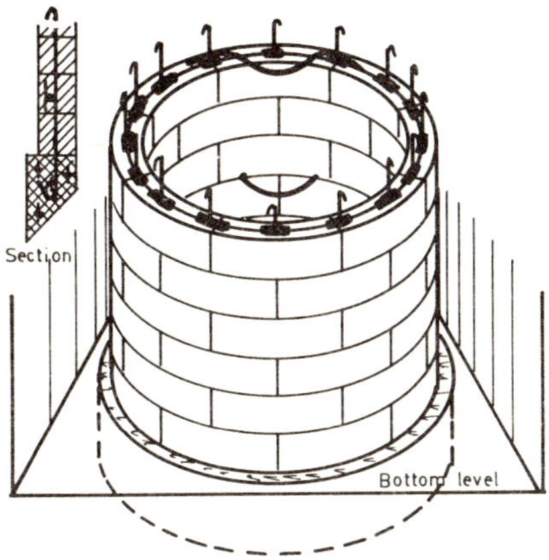

Fig. 70

d) After that place the fourth and the fifth course loosely without mortar.

Lay an iron ring upon the fifth course surrounding the iron wires.

Sprinkle the blocks and fill all the holes in the blocks with mortar. Use a stick to compact it.

e) Having filled and compacted all the holes, then cover the iron ring with mortar and press the sixth course upon it.

f) Continue to build up the well until it reaches about 1 m above ground level.

Remember that every 3rd course must have a step solidly embedded in it. In a deep well you might need a scaffolding. This is made by embedding extra steps at every 12th course just opposite the step which is there. Where two steps are situated opposite each other you can place some timbers upon them and use them as a floor for working.

g) Also remember that every 5th course must have an iron ring concreted in and that the 16 iron wires have to be connected and extended at those rings.

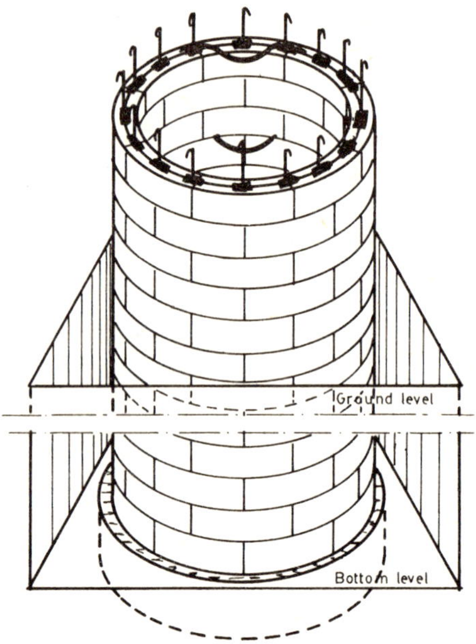

Fig. 71

4 Sinking and extension of the well

a) The well is now ready for the first sinking. A man wearing a helmet enters the well and digs out sand from the bottom of the well. He fills it into a strong bucket which is tied to a strong rope.

Another man standing at the ground level pulls up the bucket and empties its content into the corners between the well and the excavation.

Be very careful not to drop anything on the man in the bottom of the well! Slowly, as the digging proceeds, the well will sink. Stop digging when it has sunk to ground level.

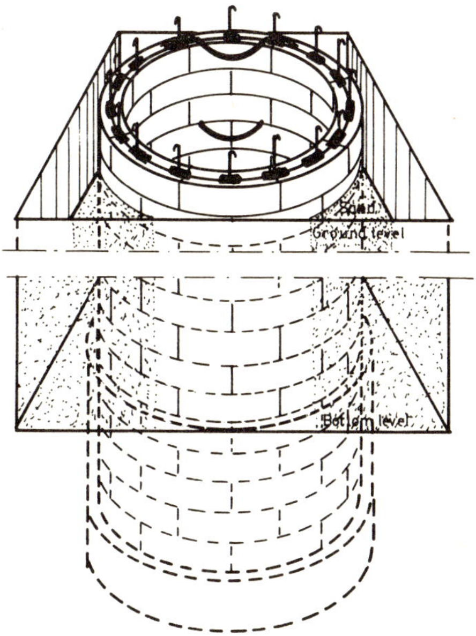

Fig. 72

b) Build another section of 5 courses onto the well.
c) Dig sand out from the bottom of the well until it has reached ground level again.

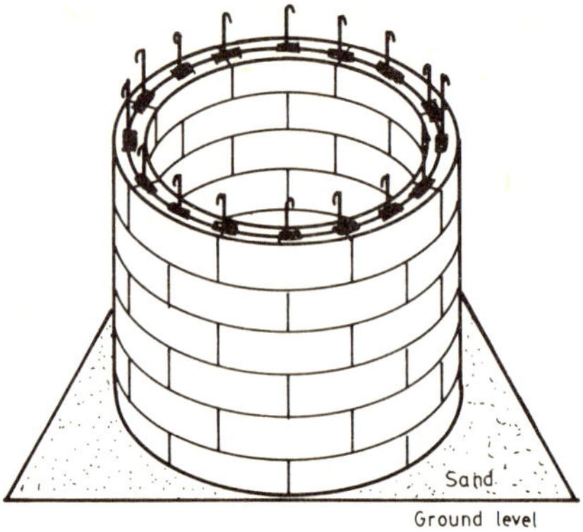

Fig. 73

d) Continue building 5 courses and digging until the well has reached the depth you want.

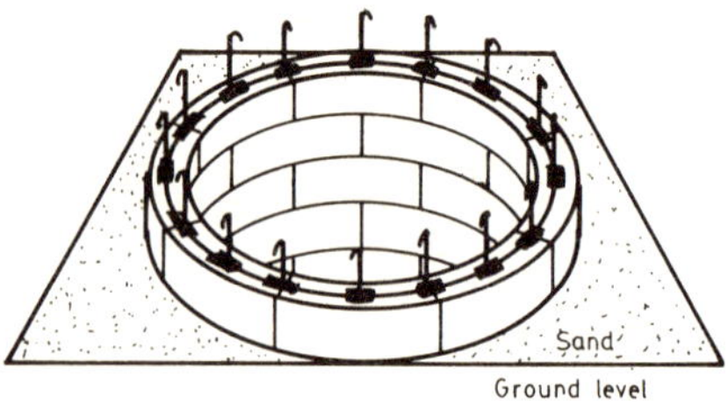

Fig. 74

5 Well head, slab and windlass

a) When you have reached the depth you want and the well has stopped sinking, it is time to build the well head.
   From one course below ground level and to the top of the well, build all blocks with mortar.
b) Build two blocks onto the last course for holding the windlass.
   Fix two pieces of hard wood in the holes of these blocks for bearings.

c) Split a wooden trunk into two halves and nail the pieces together on the windlass. Bend iron wires around the windlass and the bearings, and concrete them.
d) Plaster the edge of the well and point all joints (see pointing on page 53).
e) Concrete a slab around the well to lead away any surplus water from the well.

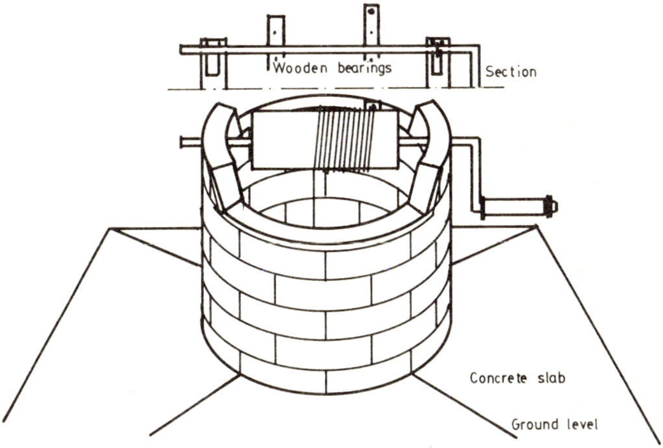

Fig. 75

Blockmaking for a sinking well

The form for making these blocks may seem a bit complicated, but hopefully these almost self-explanatory drawings and a few comments may guide you.

a) On a piece of cardboard draw the top and bottom shape of a block.
   Note that the bottom shape is 0.5 cm wider than the top one and that the patterns are exactly 1/8 of a circle.
b) Place the patterns on 6″ × 1″ timbers equidistant from the ends and edges. Draw the pattern outlines onto the timbers.

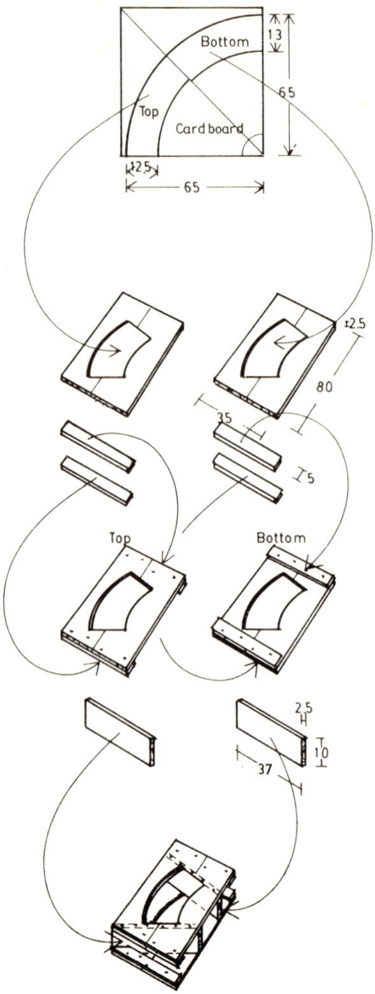

Bottom
13
65
Top
Card board
12.5
65
12.5
80
35
5
Top
Bottom
2.5
10
37

Fig. 76

c) Cut the marked areas away and nail the timbers together.
d) Nail the two sets of timber together. Be sure that the outer circle and the two ends of the cut-away areas are immediately over one another.
e) Cut two iron sheets and nail them onto the form.
f) Cut and nail two wooden pieces onto the ends of the form.

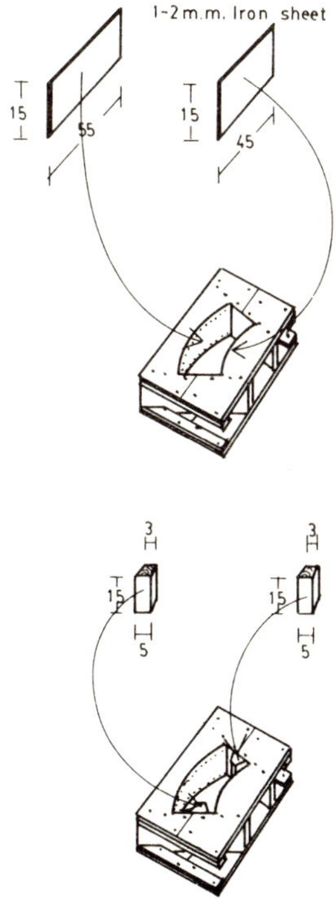

Fig. 77

g) Cut and nail together the form for making the middle hole in the blocks.

h) Cut two pieces of timber, A and B, so that they can fit into the two open sections of the form.
i) Coat the form all over with used motor oil each time before you use it.

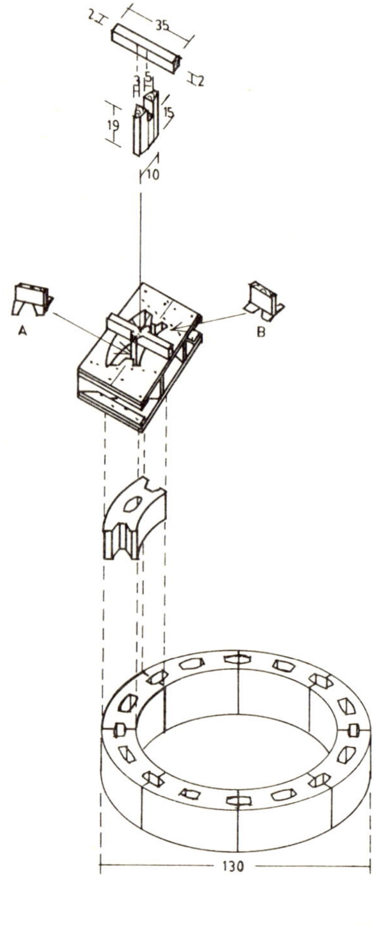

Fig. 78

j) Mix 1 bag of cement with 2½ wheelbarrows of sand, and water.
Place the form on a clean and level place.
Fill the form with concrete and compact it. Level it off flush with the surface of the form.

k) Place the two timbers, A and B, upon the concrete in the form but without touching the form. Now one man keeps these two timbers down with his hands while another man lifts the form. A block will have been formed.

l) Make 8 blocks and test that they make a circle measuring 130 cm in the outer diameter.

Recommended Reading

Subsurface dams

a) 'Handbook for Simple Water-Engineering in Kitui District'
by J P de Nooy
Diocese of Kitui
Kenya

b) 'Field Engineering for Agricultural Development'
by N W Hudson
Oxford University Press
Ely House
London W1

Storage tanks

a) 'Appropriate Technology' Vol 5 No 2
Intermediate Technology Publications Ltd
9 King Street
London WC2E 8HN

b) Granary Basket Tank
UNICEF/Kenya Government
Village Technology Unit
PO Box 44145
Nairobi

c) Large and Small Cement Jars
UNICEF/Kenya Government
Village Technology Unit
PO Box 44145
Nairobi

Wells

a) 'Where Shall We Dig the Well'
'Appropriate Technology' Vol 4 No 1
Intermediate Technology Publications Ltd
9 King Street
London WC2E 8HN

b) 'Wells Construction'
Peace Corps
Information Collection and Exchange
Washington DC 20525

c) 'Hand Dug Wells and Their Construction'
by S B Watt and W E Wood
Intermediate Technology Publications Ltd
9 King Street
London WC2E 8HN

d) 'Self-help Wells'
Irrigation and Drainage Paper No. 30
FAO
Via delle Terme di Caracalla
00100 Rome

e) 'Water Lifting Devices'
Irrigation and Drainage Series
FAO
Via delle Terme di Caracalla
00100 Rome